# STUDENT ACTIVITY WO

### HEATHER FOES
*Rock Valley College*

### KATHLEEN ALMY
*Rock Valley College*

### MICHAEL SULLIVAN, III
*Joliet Junior College*

# SULLIVAN STATISTICS SERIES
## FIFTH EDITION

# Michael Sullivan, III
*Joliet Junior College*

**PEARSON**

Boston   Columbus   Indianapolis   New York   San Francisco
Amsterdam   Cape Town   Dubai   London   Madrid   Milan   Munich   Paris   Montreal   Toronto
Delhi   Mexico City   São Paulo   Sydney   Hong Kong   Seoul   Singapore   Taipei   Tokyo

Screenshots from Minitab. Courtesy of Minitab Corporation.
Screenshot from StatCrunch. Used by permission of StatCrunch.

Reproduced by Pearson from electronic files supplied by the author.

Copyright © 2017, 2013 Pearson Education, Inc.
Publishing as Pearson, 501 Boylston Street, Boston, MA 02116.

ISBN-13: 978-0-13-411610-5
ISBN-10: 0-13-411610-0

3    17

www.pearsonhighered.com

PEARSON

# Table of Contents

The subject of statistics, like many things in life, is best learned through active participation. These activities have been designed to improve your study of statistics by helping you become a more active participant. Whether you are collecting data, discussing concepts with fellow students, or manipulating an applet, you will need to be actively involved to get the most out of these activities. The more you engage with the material, the more you will understand the concepts and the more able you will be to apply them in real life.

Even if your instructor does not assign these activities or use them in class, there are many that you can work through on your own to increase your understanding of statistical concepts.

If you are doing activities that involve the use of StatCrunch, please be sure that you are using the new version. Go to MyStatCrunch > Preferences and then select New StatCrunch before doing the activities.

# Chapter 1

# Data Collection

# Introducing Statistics Through Quotes

> Instructions: Choose a word from this word bank to complete each quote about statistics.
>
> | | | | | |
> |---|---|---|---|---|
> | numbers | lottery | experiment | truth | uncertainty |
> | illumination | certain | accuracy | assumptions | approximate |
> | averages | conclusions | suggestive | opinion | statistic |
> | statistically | probabilities | languages | make | measure |

1.  Statistics may be defined as a "body of methods for making wise decisions in the face of _____."  ~W.A. Wallis

2.  Without data, all you are is just another person with an _____.

3.  The most misleading _____ are the ones you don't even know you're making.

4.  Torture _____, and they'll confess to anything.  ~Gregg Easterbrook

5.  When you can _____what you are speaking about and express it in numbers, you know something about it; but when you cannot measure it, when you cannot express it in numbers, your knowledge is of the meager and unsatisfactory kind.  ~Lord Kelvin (British physicist)

6.  A(n) _____ answer to the right question is worth a great deal more than a precise answer to the wrong question.  ~The first golden rule of mathematics, sometimes attributed to John Wilder Tukey

7.  A knowledge of statistics is like a knowledge of foreign _____or of algebra; it may prove of use at any time under any circumstances.  ~A. L. Bowley

8.  Statistics means never having to say you're_____.  ~Unknown

9.  There are two kinds of statistics, the kind you look up and the kind you _____up.  ~Rex Stout

10. Statistics are like bikinis.  What they reveal is _____, but what they conceal is vital.  ~Aaron Levenstein

11. I abhor _____.  I like the individual case.  A man may have six meals one day and none the next, making an average of three meals per day, but that is not a good way to live.  ~Louis D. Brandeis

12. The death of one man is a tragedy.  The death of millions is a _____ _____.  ~Joseph Stalin

13.    I could prove God _____. Take the human body alone - the chances that all the functions of an individual would just happen is a statistical monstrosity. ~George Gallup

14.    The theory of _____ is at bottom nothing but common sense reduced to calculus. ~Laplace

15.    _____: A tax on people who are bad at math. ~Unknown

16.    Hysteria sells -- and _____ takes time, which could make the news stale by the time the statisticians check it out. ~Thomas Sowell

17.    I can prove anything by statistics except the _____. ~George Canning

18.    He uses statistics as a drunken man uses lampposts - for support rather than for _____. ~Andrew Lang

19.    Statistics: The only science that enables different experts using the same figures to draw different _____. ~Evan Esar

20.    All life is a(n) _____. The more experiments you make, the better. ~Ralph Waldo Emerson

# Categorizing Student Survey Data

Introduce yourself to someone sitting near you and collect the following information from him/her. Write down the information about your partner, not yourself.

1. Name:_____

2. Age: _____

3. Gender: _____

4. Birth month: _____

5. Height (in inches, to the nearest inch): _____

6. Shoe size: _____

7. Eye color: _____

8. Number of hours of TV you watched last night: _____

9. Number of credit hours you are currently taking: _____

10. Number of hours of sleep you got last night: _____

11. Time you usually go to bed: _____

12. Number of hours you worked last week: _____

13. Commute time to school in minutes: _____

14. Rating of math anxiety (none, low, med, high): _____

15. Number of songs on your digital music player: _____

16. Number of friends on Facebook:_____

17. Number of times per day that you check social media sites:_____

18. Number of tattoos:_____

19. Cell phone carrier:_____

20. Number of siblings:_____

## 6   Categorizing Student Survey Data

After collecting the data, categorize each item by filling in the chart.

| | Item | Quantitative/qualitative | Discrete/continuous | Nominal/ordinal/ interval/ratio |
|---|---|---|---|---|
| 1 | Name | | | |
| 2 | Age | | | |
| 3 | Gender | | | |
| 4 | Birth month | | | |
| 5 | Height | | | |
| 6 | Shoe size | | | |
| 7 | Eye color | | | |
| 8 | Hours of TV | | | |
| 9 | Credit hours | | | |
| 10 | Hours of sleep | | | |
| 11 | Bedtime | | | |
| 12 | Hours of work | | | |
| 13 | Commute time | | | |
| 14 | Math anxiety level | | | |
| 15 | Number of songs | | | |
| 16 | Facebook friends | | | |
| 17 | Social Media Checks | | | |
| 18 | Number of tattoos | | | |
| 19 | Cell phone carrier | | | |
| 20 | Number of siblings | | | |

# Comparing Sampling Methods

> In this activity, you will explore how different sampling techniques affect sample statistics.
>
> Consider the following question recently asked by the Gallup Organization:
>
> "In general, are you satisfied or dissatisfied with the way things are going in the country?"
>
> You will collect responses to this question using several different sampling techniques and compare the results.

**Simple Random Sample**

1.  Number the students in your class from 1 to $N$, where $N$ is the number of students in the class.

2.  Choose a sample size, $n$. _____

3.  Use a random number generator to choose $n$ students from the class to answer the question.

4.  Record the number of satisfied responses and the number of dissatisfied responses.
    Number satisfied:_____          Percent satisfied: _____
    Number dissatisfied: _____          Percent dissatisfied: _____

**Stratified Sample**

1.  Divide the students in the class by gender.  Treat each gender as a stratum.

2.  Choose a sample size, $n$, and decide how many students to sample from each stratum.
    Sample size: _____
    Number from female stratum: _____
    Number from male stratum: _____

3.  Use a random number generator to choose the sample in each stratum.

4.  Record the number of satisfied responses and the number of dissatisfied responses.
    Females:        Number satisfied:_____          Percent satisfied: _____
                    Number dissatisfied: _____          Percent dissatisfied: _____

    Males:          Number satisfied:_____          Percent satisfied: _____
                    Number dissatisfied: _____          Percent dissatisfied: _____

**Cluster Sample**

1.  Treat each row of desks in the room as a cluster.

2.  Decide how many clusters to sample.

3.  Use a random number generator to obtain a simple random sample of clusters.  Have each student in the selected clusters answer the question.

4.    Record the number of satisfied responses and the number of dissatisfied responses.
Number satisfied:_____          Percent satisfied: _____
Number dissatisfied: _____          Percent dissatisfied: _____

## Systematic Sample

1.    Number the students in your class from 1 to $N$, where $N$ is the number of students in the class.

2.    Choose an appropriate value of $k$ so you can survey every $k^{th}$ student in the class.  This can be done by computing the ratio $N/n$ and then rounding down to the nearest integer.
Value of $k$: _____

3.    Choose a starting student using a number between 1 and $k$.
Starting number, $p$:_____

4.    Obtain the systematic sample and record the number of satisfied responses and the number of dissatisfied responses.
Number satisfied:_____          Percent satisfied: _____
Number dissatisfied: _____          Percent dissatisfied: _____

## Conclusions

1.    How did the percent satisfied compare for the various sampling techniques?  Explain.

2.    Which sampling technique do you think was *most* appropriate for this situation?  Explain.

3.    Which sampling technique do you think was *least* appropriate for this situation?  Explain.

4.    What are two advantages of stratified sampling over simple random sampling?

5.    What are two practical advantages of doing a systematic sampling technique?

# Designing an Experiment

Suppose you are a toy designer who wishes to develop a new paper airplane. The design for the plane is complete, and it must be made on a 9-inch square piece of paper. However, it is unclear whether the plane should be made with newspaper-weight paper, brown-bag paper, or 24-pound printer paper. Your company wants to choose a paper that will maximize the distance that the plane will fly. This activity will guide you in designing and conducting an experiment to decide which paper should be used.

Step 1:  Identify the problem to be solved by writing the research question. Also identify the response variable.

Research question:

Response variable:

Step 2:  List at least three factors you think will affect the value of the response variable.

1.

2.

3.

4.

5.

Step 3:  Determine the number of experimental units (decide on the number of planes to make). Decide how many will be made and by whom. Consider how long it will take you to make each plane and also how long it will take to fly that many planes.

Step 4:  Determine the level of each factor. Consider the list of factors from Step 2 and decide if each should be controlled, fixed, or manipulated. Place an X in the appropriate box for each factor.

| Factor | Controlled? | Fixed? | Manipulated? |
|--------|-------------|--------|--------------|
|        |             |        |              |
|        |             |        |              |
|        |             |        |              |
|        |             |        |              |
|        |             |        |              |
|        |             |        |              |

Step 5:   Conduct the experiment.

a.        Make the planes.  Go to a website that suggests a plane design or use one of your own.

b.        As a class, conduct your experiment.  First, make sure you have considered how you will measure the flights, which unit of measurement you will use, and which measuring device you will use.

Unit of measure: _____

Measuring device: _____

c.        Prepare to collect the data.  Decide what you will need to record and set up a table to record your raw data.

d.        Conduct the experiment and record the results.

Step 6:   Test the claim.

a.          Analyze your table and compare lengths of flights against the type of paper used.

Which paper seemed to yield the longest flight?  _____

b.          Write a statement for the toy company officially recommending one of the papers.

Discuss any problems encountered during the experiment.  Is there anything you would change in the design?  In particular, are there any factors you discovered after the onset of the experiment that you believe should have been controlled that were not controlled?

What role, if any, did randomization play in your experiment?

# Chapter 2

# Organizing and Summarizing Data

# Exploring Histograms with StatCrunch

> In this activity, you will use StatCrunch to make several different histograms and explore how the starting point and bin width affect the look of the histogram. You will make a histogram that you think represents the data set well. You will also construct one histogram that you think is not a good representation and explain why it is not.

1.      Choose a data set.

   a.      In StatCrunch, open the data set that contains your class data from the Student Data Survey in chapter one or open another data set as directed by your instructor. Or, open any data set of your choice. You may explore data sets available in StatCrunch by selecting *Explore* > *Data* from the StatCrunch homepage.

   b.      Choose a variable from the data set that you would like to explore.

           Variable selected: _____

2.      Make a histogram.

   a.      With your data set open in the StatCrunch spreadsheet, open the *Histogram with Sliders Applet* by selecting *Applets* > *Histogram with Sliders*.

   b.      Choose your variable when asked to select a column.

   c.      Select *Compute!*.

3.      Explore the effect of changing the starting point for the histogram.

   a.      The *Starting point* on the histogram applet is the _____ for the frequency distribution.

   b.      Place your mouse cursor on the *Starting point* slider. Change the starting point while watching the effect this has on the histogram.

   c.      By changing the starting point, make at least two histograms that have very different shapes. Print each histogram and attach them here in the space provided. Comment on the shape that each histogram implies for the distribution of your variable.

4.    Explore the effect of changing the bin width for the histogram.

   a.    The ***Bin width*** on the histogram applet is the _____ for the frequency distribution.

   b.    Place your mouse cursor on the ***Bin width*** slider.  Change the bin width while watching the effect this has on the histogram.

   c.    By changing the bin width, make at least two histograms that have very different shapes. Print each histogram and attach them here in the space provided.  Comment on the shape that each histogram implies for the distribution of your variable.

5.    Make a histogram that represents the data well and a histogram that does not.

   a.    By changing the starting point and the bin width, make a histogram that you think represents the distribution of your selected variable well.  Print the histogram and attach it in the space provided.  Comment on the shape of the distribution of your selected variable.

   b.    By changing the starting point and the bin width, make a histogram that you think does not represent the distribution of your selected variable well.  Print the histogram and attach it in the space provided. Comment on why this is not a good histogram for your selected variable.

# Sorting Histograms by Shape

Histograms can reveal important information about the shape of a distribution for a variable. Common distribution shapes are bell-shaped, uniform, skewed right, and skewed left.

Study the histograms shown on the next pages and classify them according to their shape. In the table below, list the categories and the graphs that you included in each category. You should feel free to create categories other than those mentioned here.

Please remember that identifying the shape of a distribution from a histogram is subjective. Often a histogram will not perfectly match any shape, and you will have to use your best judgment to classify it.

| Category Name | Category Description | List of Graphs Included |
|---|---|---|
| | | |
| | | |
| | | |
| | | |
| | | |
| | | |

1.

2.

3.

4.

5.

6.

7.

8.

9.

10.

11.

12.

# Predicting Distribution Shape

Sometimes the shape of a distribution can be predicted based on the context of the data. This activity will give you an opportunity to practice predicting the shape of the distribution for data collected from your class. You will then construct histograms, identify the shape of the distribution, and judge how accurate your prediction was.

1.  Predict the shape of a distribution.

    For each of the following, identify the shape you expect for the distribution. Choose between uniform, bell-shaped, skewed right, and skewed left. Write a sentence to explain why you expect that particular shape for the data.

    a.  Random integers between 0 and 20, inclusive, generated by a calculator or statistical software

    _____

    b.  Resting pulse rates for your classmates

    _____

    c.  Size of household for your classmates

    _____

    d.  Ages of your classmates

    _____

2.  Collect the data.

    Follow your teacher's instructions to collect data for each of the variables mentioned in question 1. Store the data in your calculator or statistical software so you can make histograms from it.

3.  Make histograms.

    Construct a histogram for each data set you collected. If you are making histograms by hand, draw them on the next page. If you are using technology, print them and attach them on the next page.

4.    Analyze the shapes of the histograms.

For each data set, identify the shape of the data set based on the histogram you constructed.   Then compare this shape to the shape you predicted.  Comment on any differences between what the shape expected and the shape you found.

a.    Random integers between 0 and 20

Predicted shape:                                                    Actual shape:

Comment:

b.    Resting pulse rates

Predicted shape:                                                    Actual shape:

Comment:

c.    Size of household

Predicted shape:                                                    Actual shape:

Comment:

c.    Ages

Predicted shape:                                                    Actual shape:

Comment:

# Constructing and Comparing Graphical Representations

This activity will give you an opportunity to practice constructing graphs, either by hand or using StatCrunch. Before making a graph, you will need to consider the advantages and disadvantages of different graphical representations and choose ones that will represent the data well. After constructing your graphs, you will talk to other students and share the graphs you made.

1.    In StatCrunch, open the data set that contains your class data from the Student Data Survey in chapter one or open another data set as directed by your instructor.

Your instructor will assign you a variable to analyze from this data set.

Variable name:    _____

Qualitative or quantitative:    _____

Level of measurement:    _____

2.    Make at least two different graphs to represent your data. Draw the graph if you are making it by hand or print the graph from StatCrunch and attach it. For each graph, explain why you chose that particular type of graph and explain what it tells you about the data.

Graph #1:

Graph #2:

3.    Form a group with other students who worked on the same variable as you did. Show each other the graphs you made and discuss why you chose the particular graphs that you did.

Which graph was selected most often for your variable? Why?

Were there any graphs that were never made for your variable? If so, why were those graphs not selected?

Summarize what you learned about your variable from the graphs that were created by your group.

# Recognizing and Correcting Misleading Graphs

Unfortunately, there are many misleading and deceptive graphs published in newspapers, magazines, and in online sources. This activity will involve finding a graph that you think is misleading or deceptive and recreating the graph to give a more accurate impression.

1.      Find a graph that you think is misleading or deceptive and attach it here.  You might consider newspapers, magazines, or online sources.

2.      Discuss the difference between a graph that misleads and a graph that deceives (see section 2.4 in the textbook).  Do you think the graph you found is misleading or deceptive? Explain why.

3.        Recreate the graph to address the issue that made it misleading or deceptive.

# Chapter 3

# Numerically Summarizing Data

# Understanding Measures of Center

> In this activity you will use an applet to investigate the mean and median as measures of center and explore the resistance of these measures.

Load the **Mean Versus Median Applet** that is located at www.pearsonhighered.com/sullivanstats. Or, from StatCrunch, select **Applets > Mean/SD vs. Median/IQR**. Select the "Randomly generated" radio button. Verify the Mean and Median boxes are checked and click Compute!.

1.    Click "Reset" at the top of the applet.

    a.    Create a data set of ten observations such that the mean and median are both roughly equal to 2.

    b.    Click "Add point" and add a new observation at 9.  How does this new value affect the mean? The median?

2.    a.    Remove the single value near 9 by clicking on the point and dragging it off the number line.

    b.    Click "Add point" and add a single observation at 24. How does this new value affect the mean?  The median?

3.    Click "Reset" at the top of the applet.

    a.    Add a point at 0. Add a second point at 50. Remove these points by dragging them off the screen. (This is done to create a number line from 0 to 50).

    b.    Create a symmetric data set of six observations such that the mean and median are roughly 40.

    c.    Add a single observation at 35. How does this new value affect the mean? The median?

    d.    Grab this new point at 35 and drag it toward 0. What happens to the value of the mean? What happens to the value of the median?

4.      Click "Reset" at the top of the applet.

    a.      Add a point at 0. Add a second point at 50. Remove these points by dragging them off the screen.

    b.      Add about 25 to 30 points to create a symmetric dot plot such that the values of the mean and median are roughly 40.

    c.      Add a single observation at 35. How does this new value affect the mean? The median?

    d.      Grab this new point at 35 and drag it toward 0. What happens to the value of the mean? What happens to the value of the median?

5.      Write a paragraph that summarizes what you have learned in this activity about the mean and median. Be sure to include a discussion of the concept of resistance and the role sample size plays in resistance.

6.      Click "Reset" at the top of the applet.

    a.      Add a point at 0. Add a second point at 50. Remove these points by dragging them off the screen.

    b.      Create a data set of at least ten observations such that the mean equals the median.  What is the shape of the distribution?

    c.      Create a data set of at least ten observations such that the mean is greater than the median. What is the shape of the distribution?

d.    Create a data set of at least ten observations such that the mean is less than the median. What is the shape of the distribution?

e.    Create a data set that is skewed left, with at least 50 observations. Describe the relationship between the mean and the median.

f.    Create a data set that is skewed right, with at least 50 observations. Describe the relationship between the mean and the median.

g.    Will the relationships between the mean and median that you observed in parts e and f always hold? Can you create a data set for which they do not hold?

# Comparing Statistics to Parameters

You will begin this activity by collecting data from the population of students in your class.   Once you have collected the data, you will draw samples from the population and compare the sample mean and standard deviation to the population mean and standard deviation for your particular data.

Treat the students in your class as a population.  All the students in the class should determine their value of a variable.  You might collect values of pulse rates, heights, travel time to school, number of texts sent/received yesterday or some other variable as directed by your instructor.

1.    Number each student in the class from 1 to $N$, where $N$ is the number of students in the class.  Write each student's corresponding value of the variable next to their number.  Add more blanks as needed.

   1. _____      2. _____      3. _____      4. _____

   5. _____      6. _____      7. _____      8. _____

   9. _____      10. _____      11. _____      12. _____

   13. _____      14. _____      15. _____      16. _____

   17. _____      18. _____      19. _____      20. _____

   21. _____      22. _____      23. _____      24. _____

   25. _____      26. _____      27. _____      28. _____

   29. _____      30. _____      31. _____      32. _____

   33. _____      34. _____      35. _____      36. _____

   37. _____      38. _____      39. _____      40. _____

2.    Determine the population mean, variance, and range of the variable.

3.    Obtain a simple random sample of $n = 5$ students and record the data here.

   Determine the sample mean: _____    Does the sample mean equal the population mean?

   Determine the sample variance: _____    Does the sample variance equal the population variance?

   Determine the sample range: _____    Does the sample range equal the population range?

4.     Obtain a second simple random sample of $n = 5$ students and record the data here.

_____

       Determine the sample mean: _____ Does the sample mean equal the population mean?

       Determine the sample variance: _____ Does the sample variance equal the population variance?

       Determine the sample range: _____ Does the sample range equal the population range?

5.     Obtain a third simple random sample of $n = 5$ students and record the data here.

       Determine the sample mean: _____ Does the sample mean equal the population mean?

       Determine the sample variance: _____ Does the sample variance equal the population variance?

       Determine the sample range: _____ Does the sample range equal the population range?

6.     Are the three sample means the same?  Are the three sample variances the same?  Why or why not?

7.     The sample mean and variance are said to be *unbiased estimators* of the population mean and variance. Write a paragraph to explain what this term means.

8.     The sample range is not an unbiased estimator of the population range.  Write a paragraph to explain what this means.

# Exploring Standard Deviation

In this activity you will use an applet to investigate the standard deviation of a data set. You will investigate how to make the standard deviation smaller or larger as well as the how the standard deviation is affected by extreme values.

Load the **Standard Deviation Applet** that is located at www.pearsonhighered.com/sullivanstats. Or, from StatCrunch, select **Applets > Mean/SD vs. Median/IQR**. Select the "Randomly generated" radio button. Verify the Standard deviation box is checked and click Compute!.

1.    Click "Reset" at the top of the applet.

   a.    Click "Add point" and add a new observation at 0. Add a second point at 10. Remove these points by dragging them off the screen. (This is done to create a number line from 0 to 10).

   b.    Create a data set with ten observations so that the standard deviation is 0.5. Record a graph of your data.

   c.    By changing your previous data set or creating a new one, construct a data set so that the standard deviation is 1.0. Record a graph of your data.

   d.    By changing your previous data set or creating a new one, construct a data set so that the standard deviation is 2.0. Record a graph of your data.

   e.    Describe each of the three data sets you created in terms of their spread.

2.    Create a data set with ten observations so that the standard deviation is 0.5.  Record a graph of your data.  Grab one of the points with your mouse cursor and move it away from the rest of the data.  How does the standard deviation change?  Is the standard deviation resistant?

3.    Click "Reset" at the top of the applet.

a.    Click "Add point" and add a new observation at 0.  Add a second point at 10. Remove these points by dragging them off the screen.

b.    Create a data set with ten observations so that the standard deviation is as large as possible. What is the largest standard deviation you are able to achieve?  Record a graph of your data.

c.    Create a data set with ten observations so that the standard deviation is as small as possible. What is the smallest standard deviation you are able to achieve?  Record a graph of your data.

d.    Is it possible to create a data set with a standard deviation of 0?  If so, what does the distribution look like?

# Understanding the Standard Deviation Formula:
## Why Divide by $n - 1$?

This activity will give you an opportunity to better understand the formula for the sample standard deviation and why it involves division by $n - 1$ instead of $n$.   You will also get some practice finding sample and population standard deviations and variances, whether by hand or using technology.

Consider a family of four people, aged 8, 10, 33, and 37, as a population.

1.     Calculate the population variance and standard deviation for the ages.

Variance: _____          Standard deviation: _____

2.     Record each of the possible 16 samples of size two from this population in the second column of the chart.  Assume that random samples of size two are drawn with replacement from this population and their ages are recorded.  Instructions for completing the rest of the table continue on the next page.

| Sample | Sample Values | SD (treated as a sample) | VAR (treated as a sample) | SD (treated as a population) | VAR (treated as a population) |
|---|---|---|---|---|---|
| 1 | | | | | |
| 2 | | | | | |
| 3 | | | | | |
| 4 | | | | | |
| 5 | | | | | |
| 6 | | | | | |
| 7 | | | | | |
| 8 | | | | | |
| 9 | | | | | |
| 10 | | | | | |
| 11 | | | | | |
| 12 | | | | | |
| 13 | | | | | |
| 14 | | | | | |
| 15 | | | | | |
| 16 | | | | | |
| | | Mean = | Mean = | Mean = | Mean = |

3.    Find the variance and standard deviation for each sample (using the formula that involves division by $n - 1$). If you are using technology for the computation, the standard deviation would be labeled $s$ not sigma.

List the values in the 3rd and 4th columns of the table.

4.    Find the variance and standard deviation for each sample by treating the sample as a population (using the formula that involves division by $n$). If you are using technology for the computation, the standard deviation would be labeled sigma, not $s$.

List the values in the 5th and 6th columns of the table.

5.    Find the mean for each of the 3rd, 4th, 5th, and 6th columns. List them in the last row of the table.

6.    If the mean of a statistic (over all possible samples of the same size) is equal to the population parameter, the statistic is said to be an unbiased estimator of that parameter.

Which method (dividing by $n - 1$ or dividing by $n$ or neither) resulted in an unbiased estimate of the population variance?

Which method (dividing by $n - 1$ or dividing by $n$ or neither) resulted in an unbiased estimate of the population standard deviation?

# Scaling and Shifting Data

> This activity will help you consider how transforming data affects the mean, standard deviation, and shape of a distribution.

1.  a.  Use StatCrunch to generate 40 exam scores from a normal distribution with a mean of 75 and a standard deviation of 10. Name the column "Exam Scores."

    b.  Calculate the mean and standard deviation for your exam scores.
        Mean = _____ Standard deviation = _____

    c.  Choose one of the exam scores and calculate its z-score.

    d.  Create a histogram for the exam scores and sketch it here.

2.  Suppose the teacher plans to adjust the scores by adding ten points to each student's score.

    a.  Transform the exam score data by adding ten points to each score. Create a new column and name it "Exam Scores + 10."

    b.  Calculate the mean and standard deviation for the adjusted exam scores.

        Mean = _____ Standard deviation = _____

    c.  Calculate the z-score for the same student whose score you used in question 1.

    d.  Create a histogram for the exam scores and sketch it here.

e.      Describe how adding ten points to the exam scores affected the mean, standard deviation, and the shape of the distribution.  How did the adjustment affect the student's z-score?

3.      Suppose the teacher plans to adjust the scores by giving each student a score 20% higher than his/her original score.

a.      Transform the exam score data by multiplying each score by 1.2.  Create a new column and name it "Exam Scores x 1.20."

b.      Calculate the mean and standard deviation for the adjusted exam scores.

Mean = _____ Standard deviation = _____

c.      Calculate the z-score for the same student whose score you used in question 1.

d.      Create a histogram for the exam scores and sketch it here.

e.      Describe how adjusting the scores by multiplying by 1.20 affected the mean, standard deviation, and the shape of the distribution.  How did the adjustment affect the student's z-score?

# Matching Boxplots and Histograms

Match each histogram on this page with the appropriate boxplot on the next page.  Place the letter of the correct boxplot in the blank next to each histogram.

1. _____

2. _____

3. _____

4. _____

5. _____

6. Write a paragraph to explain how you matched the histograms and boxplots. Which characteristics of the distributions were you able to see in the graphs that helped you match them?

A.

B.

C.

D.

E.

# Chapter 4

# Describing the Relation between Two Variables

# Exploring Properties of the Linear Correlation Coefficient

> In this activity you will use an applet to create scatter diagrams and calculate the linear correlation coefficient. You will have the opportunity to observe important properties and limitations of the linear correlation coefficient.

Load the **Correlation by Eye Applet** that is located at www.pearsonhighered.com/sullivanstats. Or, from StatCrunch, select **Applets > Correlation by eye**. Select the "Randomly generated" radio button and click Compute!.

1.    Click "Reset" at the top of the applet to clear the data from the scatter diagram.

    a.    Create a scatter diagram of 12 to 15 observations with positive association. Click "Show" at the bottom of the applet to show the correlation coefficient of the data in the scatter diagram. Draw the scatter diagram (or copy it from the applet) and record the correlation coefficient below.

    b.    Move some of the observations from the scatter diagram and note how the correlation coefficient changes as the positive association strengthens and weakens.

    c.    Align the points in the scatter diagram in a straight line with positive slope. What is the value of the linear correlation coefficient?

2.    Click "Reset" at the top of the applet to clear the data from the scatter diagram.

    a.    Create a scatter diagram of 12 to 15 observations with negative association. Click "Show" at the bottom of the applet to show the correlation coefficient of the data in the scatter diagram. Draw the scatter diagram and record the correlation coefficient on the next page.

b.    Move some of the observations from the scatter diagram and note how the correlation coefficient changes as the negative association strengthens and weakens.

c.    Align the points in the scatter diagram in a straight line with negative slope.   What is the value of the linear correlation coefficient?

3.    a.    Click "Reset" at the top of the applet to clear the data from the scatter diagram.  Create a scatter diagram with no association.   What is the value of the correlation coefficient?

b.    Click "Reset" at the top of the applet to clear the data from the scatter diagram.   Create a scatter diagram in an upside-down U-shaped pattern.  Draw the scatter diagram (or copy it from the applet) and record the correlation coefficient below.

c.    What does a correlation coefficient of 0 suggest?

4.   a.   Click "Reset" at the top of the applet to clear the data from the scatter diagram.  In the lower-left corner of the applet, draw a scatter diagram of 8 to 10 observations with a correlation coefficient around 0.8.

     b.   Add another point in the upper-right corner of the applet that roughly lines up with the other points in the scatter diagram.   What is the new value of the correlation coefficient?

     c.   Move the additional point around the scatter diagram and note how the correlation coefficient changes.  Is the correlation coefficient a resistant measure?  Why or why not?

5.   a.   Click "Reset" at the top of the applet to clear the data from the scatter diagram.  Draw a scatter diagram with six points arranged vertically in a straight line.  What is the value of the correlation coefficient?  Why?

     b.   Add a seventh point to the right side of the scatter diagram  and move the point around until the correlation coefficient is approximately 0.75.

     c.   Click "Reset" at the top of the applet to clear the data from the scatter diagram.  Draw a scatter diagram with approximately seven points in a U-shaped pattern near the lower-left corner of the applet.  Add an eighth point to the scatter diagram and move it around until the correlation coefficient is approximately 0.75.  Draw the scatter diagram (or copy it from the applet).

     d.   Explain why the correlation coefficient should not be used exclusively to judge linear association without also using a scatter diagram.

# Finding a Least-Squares Regression Line

> This activity involves measuring stacks of paper to create a model that will predict the thickness of a stack of paper based on the number of pages.  You will have the opportunity to compare your model to others in the class and to use your model to make predictions.

1.    Use a ruler to measure the thickness of a single page of the textbook.  Explain why you chose the particular unit that you used.

2.    Grouping pages (one page is one sheet of paper), measure to complete the following table:

| No. of Pages | 25 | 50 | 75 | 150 | 200 | 225 |
|---|---|---|---|---|---|---|
| Thickness | | | | | | |

3.    Compute the least-squares regression line for your data in the table.

4.    Compare your data and your regression line to other students' work.  Did everyone get the same measurements and model?  Explain why or why not.

5.    Use your model to estimate the thickness of a single page of the textbook by letting $x = 1$.  Is the value you obtained reasonable?  Explain.

6.    Use your model to estimate the thickness of a single page of the textbook by interpreting the slope.  Is the value you obtained reasonable?  Explain.

7.    What, if anything, could you do to improve the model?

# Examining the Relationship
# Between Arm Length and Height

This activity uses arm length and height data collected from your class to determine a linear model. You will create a scatter diagram, find a regression equation, and then use the equation to make predictions.

1.    Each student in your class should measure the length of their forearm and their height, both as accurately as possible in centimeters. Collect the paired data (forearm length, height) and record it here.

| | | |
|---|---|---|
| _____ | _____ | _____ |
| _____ | _____ | _____ |
| _____ | _____ | _____ |
| _____ | _____ | _____ |
| _____ | _____ | _____ |
| _____ | _____ | _____ |
| _____ | _____ | _____ |
| _____ | _____ | _____ |
| _____ | _____ | _____ |
| _____ | _____ | _____ |
| _____ | _____ | _____ |

2.    Enter the data into lists on your calculator or StatCrunch or other software.

3.    Create a scatter diagram of the paired data and record it here.

4.      From the scatter diagram, does there appear to be a relationship between forearm length and height?

        If so, describe the relationship.  Does it appear to be linear?  Does it appear strong or weak?  Does there appear to be a positive or negative association between the two variables?

5.      Find the value of the linear correlation coefficient for this data and provide an interpretation.

6.      Find the least-squares regression line that relates forearm length and height.  State the equation and graph the line on the scatter diagram on the previous page.

7.      Explain what the *y*-intercept of the regression line means in this context.

8.      Explain what the slope of the regression line means in this context.

9.      a.      Use the regression equation to find the predicted height for your forearm length. _____

        b.      Calculate the difference between your actual height and the height predicted by the regression equation.    _____

        c.      What is this difference called?    _____

10.     Use the regression equation to predict the height of someone with a 6-inch long forearm.  Is your value reasonable?  Explain why or why not.

# Minimizing the Sum of the Squared Residuals

> In this activity you will use an applet to create a scatter diagram and try to determine the best-fit line by eye. You will attempt to get the sum of the squared errors for your best-fit line as low as possible and then compare your best-fit line to the regression line.

Load the **Regression by Eye Applet** that is located at www.pearsonhighered.com/sullivanstats. Or, from StatCrunch, select **Applets > Regression > by eye**. Select the "Randomly generated" radio button, check the box "Add boxes to display squared residuals," and click Compute!.

1.    Click "Reset" at the top of the applet. Create a scatter diagram of eight points with positive association.

2.    a.    Using the endpoints of the green line, attempt to fit a linear model to the data that minimizes the sum of squared residuals (or errors), SSE. Each point has a square whose side represents the residual of the point. Therefore, the area of the square is the value of the squared residual (because the area of a square equals side$^2$). So, your goal should be to make the sum of the areas of the squares as small as possible. The area of the squares is given under the heading SSE. Pay attention to the "fit" of the line versus the value of the sum of squared residuals (or errors), SSE.

      b.    Record the sum of the squared errors for the green line.
            Green (User) SSE: _____

      c.    Compare the SSE for the green line to the least-squares regression line. Record the sum of the squared errors for the regression line.

            Regression SSE: _____

      d.    How close did your SSE come to the regression SSE?

      e.    Check the "Regression" box to graph the true regression line. How close did your line come to the least-squares regression line?

      f.    Is it possible to fit a linear model to the data that makes the sum of the squared errors smaller than the sum of the squared errors for the least-squares regression line? Explain.

3.    Click "Reset" at the top of the applet. Uncheck the "Regression" box. Create a new scatter diagram by plotting eight points with negative association.

      Repeat the process from the previous exercise.

      Draw the green line until you think it is the best-fit line for the data. Observe the green SSE. Keep moving the green line to get the green SSE as low as possible. Then click "show regression line" to see how close your line is to the true regression line.

# Understanding Influential Observations

> In this activity you will use an applet to investigate the circumstances under which an observation may be influential.

Load the *Regression Influence Applet* that is located at www.pearsonhighered.com/sullivanstats.

Or, from StatCrunch, enter the data in the table below in the StatCrunch spreadsheet.

| x | y |
|---|---|
| 3 | 6 |
| 4 | 5 |
| 5 | 7 |
| 6 | 4 |
| 7 | 6 |
| 8 | 8 |
| 8 | 9 |
| 10 | 8 |
| 11 | 7 |
| 11 | 9 |
| 12 | 10 |
| 13 | 10 |
| 13 | 12 |
| 14 | 11 |

Then select *Applets* > *Regression* > *Influence*. Click the "From data table" radio button and select the *X* variable and *Y* variable columns.  Click Compute!.

1.    a.    Click "Add point." Add a point at (9, 13) by typing 9, 13 in the dialogue box. Click OK. Using your mouse, click and draw a box around the point (9, 13). You should see the point change to an x.

      b.    Check the box "Not selected" to draw the green regression line with the new point excluded. Look at the values of the intercept and slope in the table below the graph. Does it appear that the slope and/or *y*-intercept changed significantly?

2.    a.    Click "Reset" to reset the applet.

      b.    Add a point at (19, 15). Using your mouse, click and draw a box around the point (19, 15). You should see the point change to an x.

      c.    Check the box "Not selected" to draw the green regression line with the new point excluded. Look at the values of the intercept and slope in the table below the graph. Does it appear that the slope and/or *y*-intercept changed significantly?

3.    a.    Click "Reset" to reset the applet.

      b.    Add a point at (20, 8). Using your mouse, click and draw a box around the point (20, 8). You should see the point change to an x.

      c.    Check the box "Not selected" to draw the green regression line with the new point excluded. Look at the values of the intercept and slope in the table below the graph. Does it appear that the slope and/or y-intercept changed significantly?

The cases from the previous exploration are shown in the scatter diagram below. Case 1 is an outlier because the y-value is large relative to the x-value. In Case 2, both the x-value and y-value are large, but the point follows the overall pattern of the data. Finally, in Case 3, the x-value is large, and the y-value is not consistent with the pattern of the data.

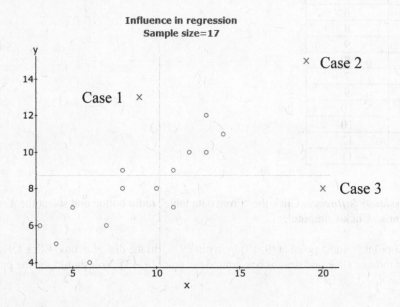

Influence is affected by two factors:

      (1)    the relative vertical position of the observation (*residuals*) and
      (2)    the relative horizontal position of the observation (*leverage*).

Leverage is a measure that depends on how much the observation's value of the explanatory variable differs from the mean value of the explanatory variable.

4.    Using these terms,

      a.    Case 1 has _____ (low/high) leverage and a _____ (small/large) residual.

      b.    Case 2 has _____ (low/high) leverage and a _____ (small/large) residual.

      c.    Case 3 has _____ (low/high) leverage and a _____ (small/large) residual.

From the previous exploration, you should conclude that observations such as Case 3 tend to be influential.

# Investigating the Coefficient of Determination

This activity will help you to understand the concept of the coefficient of determination, $R^2$.

Use the three data sets in the table below throughout the activity.

| Data Set A | | Data Set B | | Data Set C | |
|---|---|---|---|---|---|
| x | y | x | y | x | y |
| 3.6 | 8.9 | 3.1 | 8.9 | 2.8 | 8.9 |
| 8.3 | 15.0 | 9.4 | 15.0 | 8.1 | 15.0 |
| 0.5 | 4.8 | 1.2 | 4.8 | 3.0 | 4.8 |
| 1.4 | 6.0 | 1.0 | 6.0 | 8.3 | 6.0 |
| 8.2 | 14.9 | 9.0 | 14.9 | 8.2 | 14.9 |
| 5.9 | 11.9 | 5.0 | 11.9 | 1.4 | 11.9 |
| 4.3 | 9.8 | 3.4 | 9.8 | 1.0 | 9.8 |
| 8.3 | 15.0 | 7.4 | 15.0 | 7.9 | 15.0 |
| 0.3 | 4.7 | 0.1 | 4.7 | 5.9 | 4.7 |
| 6.8 | 13.0 | 7.5 | 13.0 | 5.0 | 13.0 |

1.   Notice that all three data sets have the same y-values. What is the standard deviation of y? What is the variance of y?

2.   a.   Draw a scatter diagram of each data set.

     b.   Which data set has the explanatory variable x that best explains the variability in y? Why?

3.    a.    Find the least-squares regression model between $x$ and $y$ for each data set.

Data Set A: _____

Data Set B: _____

Data Set C: _____

b.    Draw the least-squares regression line on each scatter diagram.

c.    What is the sum of squared errors for each model?

Data Set A: _____

Data Set B: _____

Data Set C: _____

d.    What does the SSE for each model suggest about the ability of the line to explain the variability in $y$?

4.    a.    Find the linear correlation coefficient between $x$ and $y$ for each data set.

Data Set A: _____

Data Set B: _____

Data Set C: _____

b.    Use the linear correlation coefficient to find the coefficient of determination, $R^2$, between $x$ and $y$ for each data set.

Data Set A: _____

Data Set B: _____

Data Set C: _____

5.    Fill in the following table.

| Data Set | Coefficient of Determination, $R^2$ | Interpretation |
|---|---|---|
| A | | _____ of the variability in $y$ is explained by the least-squares regression line. |
| B | | _____ of the variability in $y$ is explained by the least-squares regression line. |
| C | | _____ of the variability in $y$ is explained by the least-squares regression line. |

# Chapter 5

# Probability

# Demonstrating the Law of Large Numbers

In this activity you will use an applet to simulate rolling a 10-sided die and then estimate the probability of an outcome based on your results. You will explore how the probability changes as the number of trials is increased. This activity will reinforce the concept of the Law of Large Numbers and explain the empirical approach to computing probabilities.

Load the ***Rolling a Single Die Applet*** that is located at http://www.pearsonhighered.com/sullivanstats. The applet is based on rolling a single die with 10 sides.

1.  a.  Simulate rolling a 10-sided die 20 times by pressing "5 runs" four times. Be sure "Outcomes" is highlighted in gray. In the pull-down menu, select =. Change the entry in the dialogue box to 4.

    b.  What proportion of the 20 rolls results in a 4?

    c.  Based on 20 trials of the experiment, what is the probability of rolling a 4 with a 10-sided die?

2.  a.  Reset the applet. Simulate rolling a 10-sided die 50 times by pressing "5 runs" ten times. Be sure "Outcomes" is highlighted in grey. In the pull-down menu, select =. Change the entry in the dialogue box to 4.

    b.  What proportion of the 50 rolls results in a 4?

    c.  Based on 50 trials of the experiment, what is the probability of rolling a 4 with a 10-sided die?

3.  a.  Reset the applet. Simulate rolling a 10-sided die 1000 times by pressing "1000 runs" once. Be sure "Outcomes" is highlighted in grey. In the pull-down menu, select = . Change the entry in the dialogue box to 4.

    b.  What proportion of the 1000 rolls results in a 4?

    c.  Based on 1000 trials of the experiment, what is the probability of rolling a 4 with a 10-sided die?

4.    a.    With the 1000 runs from Part 3 still available, select the Convergence button.

What happens to the relative frequency with which a 4 is observed as the number of trials of the experiment increases? Recall that this illustrates the idea behind the Law of Large Numbers.

b.    Now reset the applet. Leave Convergence selected and select "1000 runs." Watch the applet carefully to see the convergence.

Is the behavior in the short run (fewer rolls of the die), the same as it was with the first 1000 runs?

What does the relative frequency converge to?

Based on this result, what is the probability of rolling a 4 with a 10-sided die?

# Finding the Probability of Getting Heads

People tend to assume that a fair penny has a 50% chance of landing on heads and a 50% chance of landing on tails.  Do you think it matters how the penny is flipped or tossed? This activity will give you the opportunity to generate data with several different methods of flipping a penny.  For each method, you will estimate the probability of getting heads and then compare your results.

1.    Toss a penny in the air, catch it with one hand, and turn it over on your other hand.  Record which side (heads or tails) is facing up when it lands.  Complete a total of 25 trials with this method, recording your result. Combine your results with the results from 3 other students to obtain a total of 100 trials.

From your experimental data, estimate the probability of getting heads with this method.

2.    Balance a penny on its edge on your desk or table, slam your hand next to it to make the penny fall. Record which side (heads or tails) is facing up when it lands.  Complete a total of 25 trials with this method, recording your result. Combine your results with the results from 3 other students to obtain a total of 100 trials.

From your experimental data, estimate the probability of getting heads with this method.

3.    Spin a penny on its edge and let it fall on its own.  Record which side (heads or tails) is facing up when it lands.  Complete a total of 25 trials with this method, recording your result. Combine your results with the results from 3 other students to obtain a total of 100 trials.

From your experimental data, estimate the probability of getting heads with this method.

4.        Compare your probabilities from the first three questions.  Were they significantly different?  Did any
          of the results surprise you?

5.        a.        Collect the probabilities of getting heads with the first method from your classmates.
                    Illustrate these probabilities with a dot plot.  Comment on the plot.

          b.        Collect the probabilities of getting heads with the second method from your classmates.
                    Illustrate these probabilities with a dot plot.  Comment on the plot.

          c.        Collect the probabilities of getting heads with the third method from your classmates.
                    Illustrate these probabilities with a dot plot.  Comment on the plot.

          d.        Comment on any differences you notice in the three plots.

# Interpreting Conditional Probabilities

The use of mammograms to screen for breast cancer for women in their forties has recently been called into question. An understanding of conditional probabilities will allow you to interpret the information accurately.

1.    Use the following information to complete the table.

If 10,000 women in their forties get a mammogram each year for a decade, then 6,432 of them will get at least one positive result and 302 of those will actually have cancer. Only 378 of the 10,000 women will develop cancer during the decade.

|  | Positive Mammogram | Negative Mammogram |  |
|---|---|---|---|
| Cancer |  |  |  |
| No Cancer |  |  |  |
|  |  |  |  |

2.    a.    What is the probability that a woman in this age group will have a positive mammogram if she has cancer?

b.    What is the probability that a woman in this age group has cancer if her mammogram was positive?

c.    Is it unusual for a woman in this age group to have cancer if she has a positive mammogram?

3.    a.    What conditional probability would tell you the false positive rate for mammography? What is the false positive rate?

b.    What conditional probability would tell you the false negative rate for mammography? What is the false negative rate?

# Calculating the Probability of Winning the Lottery

> In this activity, you will use your knowledge of counting techniques to confirm the chances of winning a particular lottery game.

The following information is available on the Illinois Lottery's website.

The **LITTLE LOTTO GAME** awards prizes to ticket holders matching two (2), three (3), four (4) or five (5) numbers from one (01) through thirty-nine (39) with the five (5) Little Lotto Game Winning Numbers randomly selected during the scheduled drawing.

## WAYS TO WIN

| Prize | Chances | AVG Prize |
|---|---|---|
| **First Prize**<br>Match all 5 numbers in any order | 1:575,757 | $100,000 starting jackpot |
| **Second Prize**<br>Match 4 of the 5 numbers in any order | 1:3,387 | $100 |
| **Third Prize**<br>Match 3 of the 5 numbers in any order | 1:103 | $10 |
| **Fourth Prize**<br>Match 2 of the 5 numbers in any order | 1:10 | $1 |
| **Overall Chances** | 1:9 | |

Source: "Ways to Win the Little Lotto". Copyright © by Illinois Department of the Lottery.

1.    Show how to calculate each of the numbers in the "Chances" column.

2.    Show how the "Overall Chances" was calculated.

3.    Do some research online to find data for a different lottery game.  Print the probability table from the website and attach it here.

4.    Show the calculations to confirm at least one of the probabilities from your chosen lottery.

# Exploring the Duplicate Birthday Problem

What is the probability of two students in your statistics class having the same birthday?  Do you think that event is likely or unlikely?  This activity will give you the opportunity to explore the probability of duplicate birthdays through data collection, a simulation, and a direct calculation.

1.  Find out if there are any duplicate birthdays (same day but not necessarily same year) in your statistics class.  One way to do this is to have all the January birthdays call out their birth date, then the February birthdays, and so on.

    How many students are in your class?  _____
    Are there any duplicate birthdays?  _____
    How likely do you think it is to have duplicate birthdays in a group of your size?  Explain.

2.  Now each student in your class will run a simulation to estimate the probability of having duplicate birthdays in a group of your size.

    a.  Use a random number generator to generate a set of birthdays (numbers 1 through 365) as large as your class size.  Sort the list to make checking for duplicates easier.

    Record the birthdays here.  Are there any duplicates?

    b.  Use the collective class data to answer the following questions.

    How many of the simulations resulted in duplicate birthdays?  _____

    How many total simulations were conducted?  _____

    What is your estimate for the probability of having duplicate birthdays in a group of your size, based on these simulations?  _____

3.    Use counting techniques to find the exact probability of having duplicate birthdays in a group of your size.

4.    Compare the estimated probability from your class simulations to the exact probability that you calculated.

5.    How many students would need to be in a class for there to be at least a 50% chance of having duplicate birthdays?

# Chapter 6

# Discrete Probability Distributions

Chapter 6

Discrete Probability Distributions

# Finding the Expected Value of a Game

> In this activity, you will play a game of chance many times and estimate the mean winnings. You will then compare your results to the expected value of the game as calculated from a probability distribution.

Consider the following game of chance. A player pays $1 and rolls a pair of fair dice. If the player rolls a sum of 2, 3, 4, 10, 11, or 12, the player loses the $1 bet. If the player rolls a sum of 5, 6, 8, or 9, there is a "push" and the player gets the $1 bet back. If the player rolls a sum of 7, the player wins $1.

1. Play the game in small groups (no money should actually change hands). Keep track of the results of the game and record the winnings for each game below (use +1 for a winner, 0 for a push, and -1 for a loser). Each group should play the game at least 25 times.

| Game | 1 | 2 | 3 | 4 | 5 | 6 | 7 | 8 | 9 | 10 |
|--------|---|---|---|---|---|---|---|---|---|----|
| Result | | | | | | | | | | |

| Game | 11 | 12 | 13 | 14 | 15 | 16 | 17 | 18 | 19 | 20 |
|--------|----|----|----|----|----|----|----|----|----|----|
| Result | | | | | | | | | | |

| Game | 21 | 22 | 23 | 24 | 25 |
|--------|----|----|----|----|----|
| Result | | | | | |

2. Combine the results of all the small groups. Determine the mean winnings.

3. Construct a probability distribution that describes the game.

4.    Determine the expected value of the game.  Is the mean obtained in #2 close to the expected value of the game?  Explain.

5.    Draw a graph similar to Figure 3 in Section 6.1 which illustrates the mean earnings versus the number of games played.

# Exploring a Binomial Distribution
## from Multiple Perspectives

This activity will help you explore a situation involving the binomial probability distribution through a physical simulation with cards, a simulation with an applet, a mathematical model, and by using statistical software.

1.    There are many estimates that exist for the percentage of the population that is left-handed. For this activity, let's assume that 10% of the population is left-handed, which is a popular estimate. We would like to build a probability model for the random variable, $X$, the number of left-handed people in a group of $n = 10$ individuals. One approach we could take is to simulate the situation.

   a.    From a standard deck of cards, select the cards Ace through 10 of a single suit (such as hearts). Let the Ace represent the left-handed individual, and the other cards represent right-handed individuals so that the probability of obtaining a left-handed individual (an Ace) from the ten individuals is 0.1.

   Select a single card from the stack.

   If the card is an Ace, record a success, otherwise, a failure.

   Replace the card and shuffle. Repeat this nine more times.

   How many successes did you obtain? _____

   b.    In part (a), we are sampling with replacement so that the outcomes of each trial are independent. Explain what this means.

   c.    Explain why the scenario laid out in part (a) qualifies as a binomial experiment.

d.    Repeat part (a) nineteen more times for a total of 20.  For each repetition of the experiment, write down how many Aces you drew (the number of successes or left-handed individuals). Then use your data to start building a probability model for $X$, the number of left-handed people in a group of 10 individuals.  Consider how many of the 20 repetitions resulted in 1 Ace, how many resulted in 2 Aces, etc.

| Experiment Number | Number of Aces |
|---|---|
| 1 | |
| 2 | |
| 3 | |
| 4 | |
| 5 | |
| 6 | |
| 7 | |
| 8 | |
| 9 | |
| 10 | |
| 11 | |
| 12 | |
| 13 | |
| 14 | |
| 15 | |
| 16 | |
| 17 | |
| 18 | |
| 19 | |
| 20 | |

| x | P(x) |
|---|---|
| 0 | |
| 1 | |
| 2 | |
| 3 | |
| 4 | |
| 5 | |
| 6 | |
| 7 | |
| 8 | |
| 9 | |
| 10 | |

e.    Combine your simulation results with those of your classmates.  Produce a dot plot of the combined data.  Use the results from your classmates to further build the probability model.

| x | P(x) |
|---|---|
| 0 | |
| 1 | |
| 2 | |
| 3 | |
| 4 | |
| 5 | |
| 6 | |
| 7 | |
| 8 | |
| 9 | |
| 10 | |

2.    Open the **Binomial Distribution Applet** at www.pearsonhighered.com/sullivanstats or from StatCrunch, open the **Coin Flipping Applet**. Let the probability of heads equal 0.1 (probability of success), and let number of coins equal 10 (the ten individuals). Press Compute!

a.    Click 1 run. How many successes (heads) did you obtain in the $n = 10$ trials of the binomial experiment?

b.    Hit Reset. Repeat the binomial experiment $N = 1000$ times by clicking 1000 runs. Based on these results, what is the probability of finding exactly two left-handed individuals in a group of ten individuals? Hint: Change number of heads to "=2."

Compare your probability to the probabilities found by your classmates. Explain why the results may differ.

3.    Rather than obtaining the probability model by drawing cards or using an applet, we could develop a mathematical model. One such model for determining the number of successes $x$ in $n$ trials of a binomial experiment where $p$ represents the probability of success is

$$P(x) = {_nC_x} \cdot p^x \cdot (1-p)^{n-x}$$

a.    We call this model the binomial probability distribution function (pdf). Use this model to find P(2), the probability of obtaining $x = 2$ left-handed people in a group of $n = 10$ people where $p = 0.1$.

b.    How does the theoretical probability compare to the probability suggested by the simulation?

c.    What would we expect to happen as the number of simulations increases if the above binomial probability model is a good model? Why?

4.    Use statistical software to build a probability model for the number of left-handed people in a group of ten people using the binomial probability distribution function. Compare the results of the simulation to the theoretical probabilities obtained from the binomial pdf.

| x | P(x) |
|---|------|
| 0 |      |
| 1 |      |
| 2 |      |
| 3 |      |
| 4 |      |
| 5 |      |
| 6 |      |
| 7 |      |
| 8 |      |
| 9 |      |
| 10 |     |

# Using Binomial Probabilities in Baseball

> This activity involves finding the probability of breaking a homerun record using simulations and the binomial probability formula. Just how unusual was it when Mark McGuire broke the homerun record in 1998?

In 1998 the baseball world was enthralled by the epic chase of Mark McGuire and Sammy Sosa to surpass the single-season homerun record of 61 set by Roger Maris in 1961.

1.    a.        Prior to his prolific 1998 season in which he shattered Roger Maris' single season homerun record of 61 by hitting 70 "round-trippers," Mark McGuire averaged 1 homerun every 11.9 at-bats. Assuming this rate of homerun hitting applied to the 1998 season, determine the probability McGuire hits a homerun during a randomly selected at-bat in 1998.

        b.        Open the **Baseball Applet** at www.pearsonhighered.com/sullivanstats or from StatCrunch, open the **Coin Flipping Applet**. Enter the probability determined in part (a) in the "Probability of heads" cell. Under the number of coins, enter 600 to represent the typical number of at-bats during the season for a starting player.

                    Run a total of 20 repetitions by clicking "5 runs" four times. What does each of these 20 repetitions represent?

                    Based on the graph, how many of the repetitions result in 62 or more homeruns (indicating Maris' record is broken)?

        c.        There are a number of players who have averaged 1 homerun every 11.9 at-bats since Maris set his record.

                    Increase the number of repetitions to 1,000 with the number of tosses at 600.

                    What does each of these 1,000 repetitions represent?

        d.        In the cell "As extreme as," enter $\geq 62$ and select "Count." This will allow us to determine the likelihood of a player hitting 62 or more homeruns in a season to break Maris' record.

2.    a.    Use the binomial probability formula to compute the exact probability of Mark McGuire breaking Maris' record over the course of a 20-season career assuming he averages 1 homerun every 11.9 at-bats.

      b.    What does this probability change to if McGuire is able to increase his homerun rate to 1 homerun every 10.8 at-bats?  This was McGuire's homerun rate in 1998.

      c.    Use the binomial probability formula to compute the exact probability of any particular player (who averages 1 homerun every 11.9 at-bats) with 600 at-bats breaking Maris' record.

3.    a.    Over the past few years, the prolific homerun hitters have been averaging 1 homerun every 13 at-bats.  Assuming the league has 10 prolific homerun hitters in any given season, what is the likelihood of McGuire's record being broken in the next 20 years?  Assume 600 at-bats per year.

            Answer using a simulation:

            Answer using the binomial probability formula:

      b.    What homerun rate would be required among the top 10 homerun hitters in order for there to be at least a 5% chance of breaking McGuire's record within the next 20 years?  Assume 600 at-bats per year.

# Designing a Game

Suppose you want to design a simple game for a fundraiser for your local school district. The game requires a player to select five balls from an urn that contains 1000 red balls and 4000 green balls.

1.  If a player selects a single ball from the urn, what is the probability of selecting a red ball?

2.  Treat the probability from part (1) as $p$, the probability of success. Assume this probability is fixed for all $n$ = 5 trials of the experiment. Explain why this assumption is reasonable.

3.  Build a probability model that describes the number of red balls selected, $X$, in $n$ = 5 trials of the experiment.

4.  You decide to charge $20 to play the game. For each game played, you would like to make $5. Decide on payouts for $x$ = 0, 1, 2, 3, 4, and 5 balls selected. Verify the expected value of the game is $5.

5.  Now, let's play the game. Open the **Urn Sampling Applet** at www.pearsonhighered.com/sullivanstats or in StatCrunch by selecting **Applets > Simulation > Urn Sampling** and conduct 1 run (play one game). How many red balls were selected? Click 1 run a second time. How many red balls were selected?

6.  Click 1000 runs at least five times (to generate at least 5000 more results of the game). Create a relative frequency distribution of the number of red balls selected based on the results of the simulation.

7.    Compare the relative frequencies from the simulation to the probability model built from the binomial probability distribution in part (3).  Do you believe the independence assumption was reasonable?

8.    Compute the mean winnings per game played based on the simulation and the payouts you decided on.

9.    Assuming the mean earnings per game are the value found in part (8).  Suppose the game is played 1000 times at the fundraiser.  How much money did the game raise for the school district?

10.    Suppose the game is only played 100 times at your fundraiser.  Explain the risks involved in making this game available.

# Chapter 7

# The Normal Probability Distribution

# Constructing Probability Distributions Involving Dice

In this activity, you will explore different distributions that arise when dealing with dice. If you simply record the number that is rolled, what kind of distribution should you expect? If you record the sum of two dice, what kind of distribution should you expect? You will run simulations to answer these questions.

1.  a.  Complete the following probability distribution for the random variable $X$ which gives the outcome when one fair die is rolled.

| $x$ | $P(x)$ |
|---|---|
| 1 | |
| 2 | |
| 3 | |
| 4 | |
| 5 | |
| 6 | |

b.  Is $X$ a continuous or discrete random variable?

How would you describe the shape of this distribution?

What is the mean?

c.  Use StatCrunch to simulate 100 rolls of a fair die.

Create a histogram of the results. Describe the shape of the distribution.

Calculate the mean from the 100 rolls. How does this value compare to the mean calculated in part b?

2.  a.  Complete the following probability distribution for the random variable $X$ which gives the sum when two fair dice are rolled.

| $x$ | $P(x)$ |
|---|---|
| 2 | |
| 3 | |
| 4 | |
| 5 | |
| 6 | |
| 7 | |
| 8 | |
| 9 | |
| 10 | |
| 11 | |
| 12 | |

b.  Is $X$ a continuous or discrete random variable?

How would you describe the shape of this distribution?

What is the mean?

c.    Use StatCrunch to simulate 100 rolls of two fair dice and find the sum for each roll.

Create a histogram of the results.  Describe the shape of the distribution.

Calculate the mean from the 100 rolls.  How does this value compare to the mean calculated in part b?  How does it compare to the mean calculated in (1b)?

3.    a.    Complete a probability distribution for the random variable $X$ which gives the average when two fair dice are rolled.

b.    Is $X$ a continuous or discrete random variable?

How would you describe the shape of this distribution?

What is the mean?

c.    Use StatCrunch to simulate 100 rolls of two fair dice.  Compute the average value of the two dice.

Create a histogram with your results.  Describe the shape of the distribution.

Calculate the expected value of the average from the 100 rolls.  How does this value compare to the mean calculated in part b?  How does it compare to the mean calculated in (1b)?

4.    a.    Calculate the probability of getting a result between 3 and 4 inclusive on one roll of a fair die.

b.    Calculate the probability of getting an average between 3 and 4 inclusive on roll of two dice.

c.    Compare the two probabilities found in parts a and b.  Which one is larger and why?

# Grading on a Curve

Suppose a professor gives a very difficult exam and wants to adjust her students' scores. This activity will allow you to explore a couple of different options for adjusted exam scores and the effects of each plan on the class scores.

1. Suppose a professor gives an exam to a class of 40 students and the scores are as follows.

| 35 | 44 | 46 | 47 | 47 | 48 | 49 | 51 | 53 | 54 |
|----|----|----|----|----|----|----|----|----|----|
| 55 | 55 | 57 | 57 | 57 | 58 | 59 | 59 | 59 | 59 |
| 60 | 60 | 60 | 60 | 60 | 62 | 62 | 62 | 64 | 68 |
| 69 | 70 | 72 | 73 | 73 | 75 | 75 | 77 | 82 | 88 |

   a. Find each of the following:

   Mean:

   Median:

   Standard deviation:

   $Z$-score for a student who originally scored 54 on the exam:

   b. Create a histogram of the scores and comment on the shape of the distribution.

2. Suppose the professor decides to adjust scores by adding 25 points to each student's score.

   a. Find each of the following after the adjustment:

   Mean:

   Median:

   Standard deviation:

   $Z$-score for a student who originally scored 54 on the exam:

   b. Create a histogram of the adjusted scores and compare it to the histogram created in #1.

3.      Suppose the professor wants the final scores to have a mean of 75 and a standard deviation of 8.

    a.      To find a student's new score, first calculate their $z$-score and then use the $z$-score along with the new mean and standard deviation to find their adjusted score. Use StatCrunch to find the adjusted score for each student.

    b.      Create a histogram of the adjusted scores and compare it to the histogram created in #1.

4.      Suppose the professor wants the scores to be curved in a way that results in the following:

        Top 10% of scores receive an A
        Bottom 10% of scores receive an F
        Scores between the 70th and 90th percentile receive a B
        Scores between the 30th and 70th percentile receive a C
        Scores between the 10th and 30th percentile receive a D

Find the range of scores that would qualify for each grade under this plan.

| Exam Scores | Letter Grade |
| --- | --- |
|  | A |
|  | B |
|  | C |
|  | D |
|  | F |

5.      Which of the three curving schemes seems the most fair to you? Explain.

# Modeling with the Normal Distribution

> This activity will allow you to explore real data with a normal distribution. You will collect data, find summary statistics, and create a histogram. Finally, you will compare percentages obtained from the actual data to those found using technology.

1.  What weight is indicated on each individual food package? _____

    Do you think all the packages will weigh that exact amount? Explain.

2.  Weigh all the individual food packages for the class and record the data here.

    | | | | |
    |---|---|---|---|
    | _____ | _____ | _____ | _____ |
    | _____ | _____ | _____ | _____ |
    | _____ | _____ | _____ | _____ |
    | _____ | _____ | _____ | _____ |
    | _____ | _____ | _____ | _____ |
    | _____ | _____ | _____ | _____ |
    | _____ | _____ | _____ | _____ |
    | _____ | _____ | _____ | _____ |
    | _____ | _____ | _____ | _____ |
    | _____ | _____ | _____ | _____ |

3.  Find the mean and standard deviation for the package weights.

4.  Create a histogram of the package weights and comment on the shape of the distribution.

5.  According to the actual data, what percent of packages had weights over _____?
    Use the normal probability commands on your calculator or software to answer this same question.
    Compare the answers you obtained from the data and technology.

6.  According to the actual data, what percent of packages had weights over _____?
    Use the normal probability commands on your calculator or software to answer this same question.
    Compare the answers you obtained from the data and technology.

# Playing Plinko

This activity will allow you to investigate the probabilities involved in the game Plinko by analyzing a quincunx on which it is based.

A quincunx is a board with pegs arranged so that the first row has one peg, the second row has two pegs, the third row has three pegs and so on. Balls are dropped from the top of the board, bounce off the pegs, and land in bins below the last row of pegs.

1.    Draw a quincunx with 10 rows of pegs. Draw the bins at the bottom and label them 0 through 10.

2.    a.    A ball will land in bin 0 if it bounces to the right 0 times on its way through the quincunx. What is the probability that a ball will land in bin 0?

       b.    A ball will land in bin 1 if it bounces to the right 1 time and to the left 9 times on its way through the quincunx. What is the probability that a ball will land in bin 1? Consider all the different paths that the ball could take and still land in bin 1.

       c.    Complete the following table by finding the probability that a ball will land in each bin.

| Bin # | 0 | 1 | 2 | 3 | 4 | 5 | 6 | 7 | 8 | 9 | 10 |
|-------|---|---|---|---|---|---|---|---|---|---|----|
| Probability | | | | | | | | | | | |

3.    Graph the probability distribution that you found in the previous question and comment on its shape.

4.    If 1,000 balls are dropped through the quincunx, how many would you expect to land in each bin?

| Bin # | 0 | 1 | 2 | 3 | 4 | 5 | 6 | 7 | 8 | 9 | 10 |
|---|---|---|---|---|---|---|---|---|---|---|---|
| Expected number of balls | | | | | | | | | | | |

5.    a.    Use the normal approximation to the binomial distribution to find the probability that a ball will land in bins 4 – 6.

b.    Use the normal approximation to the binomial distribution to find the probability that a ball will land in bins 0 – 1.

c.    Use the normal approximation to the binomial distribution to find the probability that a ball will land in bins 9 – 10.

# Analyzing Standardized Test Scores

This activity will allow you to investigate standardized test scores with a bell-shaped distribution and compare the probabilities obtained from the raw data to those from a normal distribution.

National ACT scores for 2005 were approximately normally distributed with a mean of 20.9 and a standard deviation of 4.9. The table shows the frequency of each score for the 1,186,251 students who took the exam that year.

| ACT Composite Score | Frequency | Cumulative Percentage |
|---|---|---|
| 36 | 193 | 99 |
| 35 | 1617 | 99 |
| 34 | 3729 | 99 |
| 33 | 6752 | 99 |
| 32 | 10359 | 99 |
| 31 | 15401 | 98 |
| 30 | 21725 | 97 |
| 29 | 27109 | 95 |
| 28 | 35992 | 93 |
| 27 | 44428 | 90 |
| 26 | 52811 | 86 |
| 25 | 62669 | 81 |
| 24 | 70594 | 76 |
| 23 | 77953 | 70 |
| 22 | 83849 | 64 |
| 21 | 89908 | 57 |
| 20 | 91333 | 49 |
| 19 | 89073 | 41 |
| 18 | 87315 | 34 |
| 17 | 79103 | 26 |
| 16 | 70716 | 20 |
| 15 | 59275 | 14 |
| 14 | 45796 | 9 |
| 13 | 31658 | 5 |
| 12 | 18095 | 2 |
| 11 | 6586 | 1 |
| 10 | 1594 | 1 |
| 9 | 420 | 1 |
| 8 | 138 | 1 |
| 7 | 41 | 1 |
| 6 | 12 | 1 |
| 5 | 7 | 1 |
| 4 | 0 | 1 |
| 3 | 0 | 1 |
| 2 | 0 | 1 |
| 1 | 0 | 1 |

1.    Construct a graph of the frequencies for each score on the ACT exam.  Describe the shape of the distribution of scores.

2.    a.    Using the raw data contained in the table, what percentage of students scored between 15 and 25 (inclusive) on the test?

      b.    Using the normal distribution, what percentage of students would you estimate scored between 15 and 25 (inclusive) on the test?

3.    a.    Using the raw data contained in the table, what percentage of students scored below 18?

      b.    Using the normal distribution, what percentage of students would you estimate scored below 18?

4.    a.    Using the raw data contained in the table, what percentage of students scored above 28?

      b.    Using the normal distribution, what percentage of students would you estimate scored above 28?

5.    a.    Using the raw data contained in the table, what score is in the 75th percentile?

      b.    Using the normal distribution, what score would you expect to be in the 75th percentile?

6.    If a college program requires a minimum score of 24 for placement, what percent of students would **not** qualify?  Answer using the raw data and using the normal distribution.

# Chapter 8

# Sampling Distributions

# Creating a Sampling Distribution for the Mean

> In this activity, you will randomly select a small group of your classmates and then determine all the possible samples from that population. You will find the mean for each sample and then construct a sampling distribution for the mean. The goal of the activity is to better understand the shape, center, and spread of the sampling distribution for the mean.

1.    Randomly select six students from the class to treat as a population. Choose a quantitative variable (such as pulse rate, age, or number of siblings) to use for this activity, and gather the data appropriately. Share the data with the class.

   Compute the mean for this population.

2.    a.    Work with your group to list all the samples of size $n = 2$, $n = 3$, $n = 4$, or $n = 5$, as assigned by your instructor.

   Collect the results from the other groups to complete the information for all sample sizes.

   Hint: The sample of size $n = 2$ should have $_6C_2 = 15$ possible samples.

   Samples of size $n = 2$                    Samples of size $n = 3$

   Samples of size $n = 4$                    Samples of size $n = 5$

   b.    Compute the sample mean for each sample listed for your group's sample size in part (a).

   Record the means next to the samples.

   Collect the results from the other groups to complete the information for all sample sizes.

    c.      Form the probability distribution for the sample mean by making a table for each sample size and listing all the possible sample means with their probabilities.

3.        Verify for each probability distribution in (2c) that $\mu_{\bar{x}} = \mu_x$, that is, the mean of the sample means is the population mean.

4.        Draw the probability histogram for your group on the board.  Compare the spread in each probability distribution based on the probability histogram.  What does this result imply about the standard deviation of the sample mean?

# Analyzing the Variability in Sample Means

> The purpose of this activity is to explore the amount of dispersion in values of a variable for individuals versus the variability in means.

1.    Each student should determine his or her at-rest pulse rate.  Be sure to agree as a class on the method for determining the pulse rate.  Share the data with the class.

Treat the students in the class as a sample of all students at the school.

Determine the sample standard deviation of the at-rest pulse rates.

2.    a.    Obtain a simple random sample of $n = 5$ students from the class.  Compute the mean pulse rate of the students in the sample.  Repeat this five more times, for a total of six sample means.

| Sample | Data Values in Sample | | | | | Sample Mean |
|--------|--|--|--|--|--|------|
| 1 | | | | | | |
| 2 | | | | | | |
| 3 | | | | | | |
| 4 | | | | | | |
| 5 | | | | | | |
| 6 | | | | | | |

   b.    Compute the sample standard deviation of the mean pulse rates found in part (a).

   c.    Compare the sample standard deviation of the mean pulse rates from part (b) to the sample standard deviation of the pulse rates from #1.

   Which has more dispersion, the individual pulse rates in the class, or the mean pulse rates of the five students?  Why?

3.    If a simple random sample of $n = 10$ students from class was used instead, would the standard deviation of the mean pulse rates be higher or lower than the standard deviation of the mean pulse rates with $n = 5$ students?  Explain.

# Simulating IQ Scores

This activity will involve simulating data to represent IQ scores. After generating large samples of IQ scores, you will construct a sampling distribution for the mean and find its mean and standard deviation. You will also compare proportions calculated from the sampling distributions to those predicted by the normal distribution.

1.  a.  Use StatCrunch, MINITAB, Excel, or some other statistical software to obtain 500 random samples of size $n = 10$ from the population of IQ scores with a mean of 100 and a standard deviation of 15.

    b.  Compute the sample mean for each of the 500 samples. Provide the sample mean for Sample 1 and Sample 2 here.

    Mean from sample 1:      _____

    Mean from sample 2:      _____

2.  a.  What do you expect the mean and standard deviation of the sampling distribution of the mean to be?

    b.  Draw a histogram of the 500 sample means. If possible, draw the theoretical normal model on the histogram.

c.      Determine the mean and standard deviation of the 500 sample means.  Are these values close to what was expected in part (a)?

d.      What proportion of the 500 random samples resulted in a sample mean IQ greater than 108?

Based on the normal model, what proportion of random samples of size $n = 10$ would we expect to result in a sample mean greater than 108?

Is the theoretical proportion obtained from the model close to the proportion from the simulation?  Explain.

# Sampling from Normal and Non-Normal Populations

> In this activity, you will use an applet to simulate data from three different distributions. In each case, you will describe the center, spread, and shape of the distribution of the sample mean.

1.  a.   Load the ***Sampling Distribution Bell Shaped Applet*** from www.pearsonhighered.com/sullivanstats or from StatCrunch. Set the applet so that the population is bell-shaped. Decide on a mean and standard deviation and make a note of them here.

    b.   Obtain 1,000 random samples of size $n = 5$, $n = 10$, and $n = 20$.

    c.   Describe the distribution of the sample means based on the results of the applet. That is, describe the center (mean), spread (standard deviation), and shape of the distribution.

|  | $n = 5$ | $n = 10$ | $n = 20$ |
|---|---|---|---|
| mean | | | |
| standard deviation | | | |
| shape | | | |

    d.   Compare the distributions for the various sample sizes.

    e    What role, if any, does the sample size play in the sampling distribution of the sample mean?

2.  a.   Load the ***Sampling Distribution Uniform Applet*** from www.pearsonhighered.com/sullivanstats or from StatCrunch. Set the applet so that the population is uniform.

    b.   Obtain 1,000 random samples of size $n = 2$, $n = 5$, $n = 10$, $n = 20$, and $n = 30$.

    You may stop generating random samples once the distribution of the sample mean is approximately normal.

    c.   Record the mean and the standard deviation of the distribution of sample means for each sample size.

|  | $n = 2$ | $n = 5$ | $n = 10$ | $n = 20$ | $n = 30$ |
|---|---|---|---|---|---|
| mean | | | | | |
| standard deviation | | | | | |

d.    For which sample size is the distribution of sample means approximately normal?

e.    What is the mean of each distribution (regardless of shape)?

f.    What is the standard deviation of each distribution (regardless of shape)?

g.    How does the size of the sample affect the standard deviation of the sample means?

3.    a.    Draw a skewed distribution on the applet by holding the left mouse button down and dragging the mouse over the distribution.

b.    Obtain 1,000 random samples of size $n = 2$, $n = 5$, $n = 10$, $n = 20$, and $n = 30$.

You may stop generating random samples once the distribution of the sample mean is approximately normal.

c.    Record the mean and the standard deviation of the distribution of sample means for each sample size.

|  | $n = 2$ | $n = 5$ | $n = 10$ | $n = 20$ | $n = 30$ |
|---|---|---|---|---|---|
| mean |  |  |  |  |  |
| standard deviation |  |  |  |  |  |

d.    For which sample size is the distribution of sample means approximately normal?

e.    What is the mean of each distribution (regardless of shape)?

f.    What is the standard deviation of each distribution (regardless of shape)?

g.    How does the size of the sample affect the standard deviation of the sample means?

# Creating a Sampling Distribution for a Proportion

> In this activity, you will randomly select a small group of your classmates and then determine all the possible samples from that population. You will find a proportion for each sample and then construct a sampling distribution for the proportion. The goal of the activity is to better understand the shape, center, and spread of the sampling distribution for the proportion.

This activity will use the data that was collected in the first Chapter 8 activity entitled *Creating a Sampling Distribution for the Mean*. If you did that activity, use the data from it, along with the samples created from it, for the first two parts of this activity.

1.      Randomly select six students from the class to treat as a population. Choose a quantitative variable (such as pulse rate, age, or number of siblings) to use for this activity, and gather the data appropriately. Share the data with the class.

   Compute a proportion for this population, as directed by your instructor.

2.      a.      Work with your group to list all the samples of size $n = 2$, $n = 3$, $n = 4$, or $n = 5$, as assigned by your instructor.

   Collect the results from the other groups to complete the information for all sample sizes.

   Hint: The sample of size $n = 2$ should have $_6C_2 = 15$ possible samples.

   Samples of size $n = 2$                    Samples of size $n = 3$

   Samples of size $n = 4$                    Samples of size $n = 5$

   b.      Compute the sample proportion for each sample listed for your group's sample size in part (a).

   Record the proportions next to the samples.

   Collect the results from the other groups to complete the information for all sample sizes.

    c.       Form the probability distribution for the sample proportion by making a table for each sample size and listing all the possible sample proportions with their probabilities.

3.       Verify for each probability distribution in (2c) that the mean of the sample proportions is the population proportion.

4.       Draw the probability histogram for your group on the board.  Compare the spread in each probability distribution based on the probability histogram.  What does this result imply about the standard deviation of the sample proportion?

# Describing the Distribution of the Sample Proportion

In this activity, you will use an applet to simulate data from a binary distribution. You will describe the center, spread, and shape of the distribution of the sample proportion for various sample sizes.

1.  a.  Load the **Sampling Distribution Binary Applet** from www.pearsonhighered.com/sullivanstats or from StatCrunch. Set the applet so that the population is binary with a probability of success equal to 0.25.

    b.  Obtain 1,000 random samples of size $n = 5$, $n = 25$, and $n = 75$.

    c.  Describe the distribution of the sample proportions based on the results of the applet. That is, describe the center (mean), spread (standard deviation), and shape of the distribution.

|  | $n = 5$ | $n = 25$ | $n = 75$ |
|---|---|---|---|
| mean |  |  |  |
| standard deviation |  |  |  |
| shape |  |  |  |

    d.  Compare the distributions for the various sample sizes.

    e.  What is the mean of the distribution in all three cases (approximately)?

    f.  What role does sample size play in the standard deviation?

# Chapter 9

## Estimating the Value of a Parameter

# Exploring the Effects of Confidence Level, Sample Size, and Shape I

In this activity, you will use an applet to generate confidence intervals for the population proportion. You will use the results to understand what is meant by the confidence level and the role of sample size.

Open the **Confidence Intervals for a Proportion with p = 0.3 Applet** from www.pearsonhighered.com/sullivanstats or from StatCrunch by selecting **Applets > Confidence intervals > for a proportion**.

1.    Exploring the role of level of confidence:

    a.    Construct 1,000 confidence intervals with $n = 100$, $p = 0.3$. What proportion of the 95% confidence intervals included the population proportion, 0.3?

        Construct another 1,000 confidence intervals with $n = 100$, $p = 0.3$. What proportion of the 95% confidence intervals included the population proportion, 0.3?

        Did the same proportion of intervals include the population proportion each time? What proportion did you expect to include the population proportion?

    b.    Construct 1,000 confidence intervals with $n = 100$, $p = 0.3$. What proportion of the 99% confidence intervals included the population proportion, 0.3?

        Construct another 1,000 confidence intervals with $n = 100$, $p = 0.3$. What proportion of the 99% confidence intervals included the population proportion, 0.3?

        Did the same proportion of intervals include the population proportion each time? What proportion did you expect to include the population proportion in each part?

2.    Exploring the role of sample size:

    a.    Construct 1,000 confidence intervals with $n = 10$, $p = 0.3$. What proportion of the 95% confidence intervals included the population proportion, 0.3?

b.       Construct 1,000 confidence intervals with $n = 40$, $p = 0.3$.  What proportion of the 95% confidence intervals included the population proportion, 0.3?

c.       Construct 1,000 confidence intervals with $n = 100$, $p = 0.3$.  What proportion of the 95% confidence intervals included the population proportion, 0.3?

d.       Did the same proportion of intervals include the population proportion in parts (a), (b), and (c)?

What proportion did you expect to include the population proportion in each part?

What happens to the proportion of intervals that include the population proportion as the sample size increases?  Why?

3.       Further exploration:

a.       Construct 100 confidence intervals with $p = 0.3$, level of confidence $= 0.95$, and sample size $= 100$.

Select one of the intervals (denoted with a red line) that does not include the population proportion that is to the left of the population proportion, 0.3.  If there are no such intervals, repeat until there are. Scroll the mouse cursor over the interval.  In the popup window, notice the value of the sample proportion.

Determine the number of standard errors the sample proportion is from the population proportion, 0.3, by computing  $z = \dfrac{\hat{p} - 0.3}{\sqrt{\dfrac{\hat{p}(1 - \hat{p})}{100}}}$ . Remember, the standard error is

$$\sigma_{\hat{p}} = \sqrt{\dfrac{\hat{p}(1 - \hat{p})}{n}} \, .$$

b.   From the same 100 confidence intervals constructed in part a, select one of the intervals that does not include the population proportion (red line) that is to the right of the population proportion.  In the pop-up window, notice the value of the sample proportion.  Determine the number of standard errors the sample proportion is from the population proportion, 0.3.

c.   Are each of the sample proportions from parts a and b more than 1.96 standard errors from the population proportion?   Explain why any sample proportion that is more than 1.96 standard errors from the population proportion will result in an interval that does not include the population proportion.

# Constructing a Confidence Interval with M&M's

Are the colors evenly distributed in a bag of M&M's? Is the company reporting accurate color percentages for their product? This activity will guide you through constructing a confidence interval for the proportion of M&M's in a particular color. You will compare your confidence interval to those of your classmates and see what percent of them actually contain the proportion reported by the manufacturer.

You have long suspected that the colors are not evenly distributed in a bag of M&M's. Out of curiosity, you decide to construct a confidence interval for the proportion of orange M&M's in a bag of peanut M&M's.

1.      a.      Treat your bag of peanut M&M's as a simple random sample.

        b.      Count the number of orange M&M's and the total number of M&M's in your bag. Then determine the proportion of your M&M's that are orange.

                Number of orange M&M's:            _____

                Total number of M&M's:              _____

                Proportion of orange M&M's:        _____

        c.      Verify that the requirements for constructing a confidence interval for the proportion are satisfied. If the conditions are not met, what should you do?

        d.      Share your data with the rest of the class, according to your instructor's directions.

2.      Construct a 95% confidence interval for the proportion of orange M&M's. Be sure to include an interpretation of your interval.

3.      On the board, construct a scale and draw your confidence interval as a line segment, along with those of your classmates.

4.    a.    According to Mars, Inc., peanut M&M's are _____ orange.

      b.    What percent of the class's confidence intervals contained the true proportion of orange
            M&M's? _____

            What percent of the confidence intervals did you expect to contain the true proportion?
            Explain.

5.    a.    If you had a larger sample, how would you expect the confidence interval to be affected?

      b.    Pool the class data and assume that all the M&M's constitute one large sample.

            Number of orange M&M's:          _____

            Total number of M&M's:           _____

            Proportion of orange M&M's:      _____

      c.    Construct a 95% confidence interval for the proportion of orange M&M's based on the pooled
            class data.  Be sure to include an interpretation of your interval.

      d.    Does this confidence interval contain the true proportion of orange M&M's?

      e.    How does the margin of error for this confidence interval compare to the margin of error for
            your individual confidence interval?

# Constructing a Confidence Interval for Average Temperature

> In this activity, you will collect a random sample of daily high temperatures for a specific city for a specific month and use it to build a confidence interval for the mean high temperature for that month in the city.

1.  Go to WeatherUnderground.com or another weather website. Search for weather data for a particular city.

    Check which years of data are available for your chosen city.  Use a random number generator to randomly select a year in this range and a day of the month. Then record the high temperature for your city on that day. Repeat this process until you have a random sample of 30 high temperatures. Record the city, date,s and temperatures below.

    City:_____

| Date | High Temperature | Date | High Temperature |
|------|------------------|------|------------------|
|      |                  |      |                  |
|      |                  |      |                  |
|      |                  |      |                  |
|      |                  |      |                  |
|      |                  |      |                  |
|      |                  |      |                  |
|      |                  |      |                  |
|      |                  |      |                  |
|      |                  |      |                  |
|      |                  |      |                  |
|      |                  |      |                  |
|      |                  |      |                  |
|      |                  |      |                  |
|      |                  |      |                  |
|      |                  |      |                  |
|      |                  |      |                  |

2.  Find a 95% confidence interval for the mean high temperature for your city for your chosen month. Write your interval in the form *mean +/- margin of error* and include a sentence to interpret your interval.

# Exploring the Effects of Confidence Level, Sample Size, and Shape II

In this activity, you will use an applet to generate confidence intervals for the population mean. You will use the results to understand what is meant by the confidence level and the role of sample size. You will also explore how the shape of the original distribution affects the confidence intervals.

Open the **Confidence Intervals for a Mean (Normal) Applet** from www.pearsonhighered.com/sullivanstats or from StatCrunch by selecting **Applets > Confidence intervals > for a mean**.

1.    Exploring the role of level of confidence:

a.    Set the shape to normal with mean = 50 and standard deviation = 10.

b.    Simulate at least 1,000 simple random samples from this population with $n = 10$.
For 95% confidence, what proportion of the confidence intervals included the population mean?

c.    Simulate another 1,000 simple random samples from this population with $n = 10$.
For 95% confidence, what proportion of the confidence intervals included the population mean?

d.    Did the same proportion of intervals include the population mean in parts (b) and (c)?

Explain what this result implies.

What proportion did you expect to include the population mean? Explain any discrepancy.

2.    Exploring the role of sample size:

a.    Set the shape to normal with mean = 50 and standard deviation = 10.

b.    Simulate at least 1,000 simple random samples from this population with $n = 10$.
For 95% confidence, what proportion of the confidence intervals included the population mean?

c.    Simulate 1,000 simple random samples from this population with $n = 20$.
For 95% confidence, what proportion of the confidence intervals included the population mean?

d.    Simulate 1,000 simple random samples from this population with $n = 40$.
For 95% confidence, what proportion of the confidence intervals included the population mean?

e.    Compare the width of the intervals for parts (b) – (d).

Which intervals have the larger width?  Explain your response.

3.    Exploring the role of shape:
Open the **Confidence Intervals for a Mean (Skewed) Applet** from
www.pearsonhighered.com/sullivanstats or from StatCrunch by selecting **Applets > Confidence intervals > for a mean**.

a.    Set the shape to skewed right with $n = 5$, mean = 50, and standard deviation = 10.

Simulate at least 1,000 simple random samples from this population.
For 95% confidence, what proportion of the confidence intervals included the population mean?

b.    Set the shape to skewed right with $n = 10$, mean = 50, and standard deviation = 10.

Simulate at least 1,000 simple random samples from this population.
For 95% confidence, what proportion of the confidence intervals included the population mean?

What proportion of intervals would you expect to include the population mean?

Is the proportion of simulated intervals that included the population mean closer to what is expected than what was observed in part (a)?  Explain.

c.    Simulate obtaining random samples with a variety of sample sizes.  In your judgment, what sample size is needed in order for the proportion of intervals that include the mean to be equal to the proportion suggested by the level of confidence?

Be sure to clearly describe your approach that led to your conclusion.

# Constructing a Confidence Interval
# from a Non-Normal Distribution

The exponential probability distribution is a probability distribution that can be used to model waiting time in line or the lifetime of electronic components. Its density function is given by the formula

$f(x) = \frac{1}{\mu} e^{-x/\mu}$ where $x \geq 0$. Its shape is skewed right. Suppose the wait-time in a line can be modeled by the exponential distribution with $\mu = 5$ minutes.

1.  Simulate obtaining a random sample of 200 individuals waiting in line where the mean wait-time is 5 minutes.

    Draw a histogram of the data and comment on its shape.

2.  Use StatCrunch, Minitab, or some other statistical software to simulate obtaining 100 simple random samples of size $n = 10$ from the population of 200 individuals waiting in line.

3.  a.  Determine the mean and standard deviation of the first simulated sample.

        Mean = _____
        Standard deviation = _____

    b.  By hand, construct a 95% confidence interval for the mean wait time based on the first sample. Does your interval include the mean of 5 minutes?

4.    a.    Determine 90% *t*-intervals for each of the 100 samples using the statistical software.

      b.    How many of the intervals do you expect to include the population mean?

      c.    How many of the intervals actually do include the population mean?

      d.    Explain what your results suggest.

5.    Repeat #2 – 4 for random samples of size $n = 20$ and $n = 30$.

      Write a report detailing the impact of sampling from a non-normal distribution on the actual proportion of intervals that include the population mean versus the expected proportion.

# Constructing a Confidence Interval for Die Rolls

How likely are you to roll a 1 with a fair die?  How many rolls, on average, should it take to roll a 1?  This activity will help you use confidence intervals to explore this question.

1. Either as an individual or in a small group, roll a die until a "1" is observed.  Repeat this process a total of 30 times and record the number of rolls required to get a "1" in each of the 30 trials of the experiment.

| Trial # | Number of rolls to get a 1 | Trial # | Number of rolls to get a 1 | Trial # | Number of rolls to get a 1 |
|---|---|---|---|---|---|
| 1 | | 11 | | 21 | |
| 2 | | 12 | | 22 | |
| 3 | | 13 | | 23 | |
| 4 | | 14 | | 24 | |
| 5 | | 15 | | 25 | |
| 6 | | 16 | | 26 | |
| 7 | | 17 | | 27 | |
| 8 | | 18 | | 28 | |
| 9 | | 19 | | 29 | |
| 10 | | 20 | | 30 | |

2. Use your results to obtain a point estimate of the mean number of rolls of a die required to obtain a "1."

3. Draw a histogram of your data.  What is the shape of the histogram?

4. Construct a 90% confidence interval for the mean number of rolls required to obtain a "1."

5. The population mean number of rolls to obtain a "1" is 6.  Does your interval include 6? _____

   What proportion of the intervals in the class included 6? _____

   What proportion of the intervals in the class did you expect to include 6? _____

# Take Me Out to the Ball Game

> In this activity we explore the population of all homeruns hit during the 2014 major league baseball season. In particular, we consider the distance each ball traveled as it left the park.

In this activity we will explore the population of all homeruns hit during the 2014 major league baseball season. In particular, we will consider the distance each ball traveled as it left the park. Open the data set Homerun at www.pearsonhighered.com/sullivanstats/HomeRun2014. The column "TrueDist" represents the distance, in feet, that each homerun traveled.

1.     Determine the population mean distance of a homerun in 2014. Determine the population standard deviation distance of a homerun in 2014.

2.     a.     Draw a relative frequency histogram of the variable "TrueDist."

       b.     Describe the shape of the distribution. Does it seem reasonable to conclude the variable "TrueDist" is a normal random variable?

       c.     What proportion of the homeruns were more than 450 feet?

3.     a.     Obtain a simple random sample of size $n = 10$ homeruns from this population. List the distance each of the 10 homeruns traveled.

       b.     Determine the sample mean and sample standard deviation for the sample data found in part 3a.

       c.     Construct a 95% confidence interval for the mean distance of a homerun hit in 2014 based on the sample data from part 3a.

d.      Does your confidence interval include the population mean?

4.    a.      Obtain a second simple random sample of size $n = 10$ homeruns from the population of all homeruns hit in 2014.

b.      Determine the sample mean and sample standard deviation for the sample data found in part 4a.

c.      Construct a 95% confidence interval for the mean distance of a homerun hit in 2014 based on the sample data from part 4a.

d.      Explain why the confidence interval from part 3c is different from that of part 4c.

5.    a.      Construct a 99% confidence interval for the random sample found in part 3a.

b.      Is the confidence interval narrower or wider?  Why?

6.    a.      Obtain a simple random sample of size $n = 20$ homeruns from this population.

b.      Construct a 95% confidence interval for the mean distance of a homerun based on the sample data from part 6a.  Is the interval narrower or wider than the interval from part 3c.?  Why?

7.      As a class, determine the proportion of all the intervals (based on the samples from parts 3 and 4) that include the population proportion.  Explain what this result suggests.

# Finding a Bootstrap Confidence Interval

> Bootstrapping is sampling with replacement from a sample that is representative of the population whose parameters we wish to estimate. You will obtain many random samples with replacement from the sample data and compute the mean of each random sample. For a 95% confidence interval, you will find the cutoff points for the middle 95% of the sample means. That is, you will find the 2.5[th] and 97.5[th] percentiles. These cutoff points represent the lower and upper bounds of the confidence interval.

The data in the table represent survival time (in years) of a random sample of eight patients with a rare disease.

| 1.0 | 0.7 | 0.6 | 0.5 |
|-----|-----|-----|-----|
| 0.3 | 1.4 | 3.1 | 0.2 |

1.     Verify that 3.1 is an outlier. Use this result to explain why a 95% confidence interval for the mean survival time using Student's $t$-distribution cannot be determined.

2.     Open an Excel spreadsheet and enter the sample data in rows 1 and 2, columns A through D.

   a.     To sample with replacement, enter the following command in cell A4:

   =index($a$1:$d$2,rows($a$1:$d$2)*rand()+1,columns($a$1:$d$2)*rand()+1)

   Copy the contents in cell A4 to cells B4 through H4 (because the size of the original sample is $n = 8$, each of our resamples will also have n = 8 observations). This creates the first random sample with replacement of size $n = 8$. List the observations in the first resample below.

   b.     Copy the contents in cell A4 through H4 to cells A5:H5 through A203:H203. You now have 200 random samples with replacement of $n = 8$. List the observations in the second and third resample below.

   c.     Determine the mean of each of the 200 random samples by entering

   = average (A4:H4)

   into cell J4. Copy cell J4 into cells J5 through J203. You now have 200 sample means. List the first three sample means below.

3.    Draw a histogram of the 200 sample means.  Do you believe the distribution of the sample mean is approximately normal?  Explain.

4.    We want to find the middle $(1-\alpha)100\%$ of the data.

a.    Enter the level of confidence (as a decimal) into cell L1.

b.    In cell L2, enter $=1-L1$ to get the value of $\alpha$.  Then enter the following into cell L3 to determine the number of sample means:

= count(J4:J203)

c.    Find the lower bound by entering the following into cell L4:

=small(J4:J203,L2/2*L3)

What is the lower bound?

d.    Find the upper bound by entering the following into cell L5:

=small(J4:J203,(1-L2/2)*L3)

What is the upper bound?

e.    Interpret the confidence interval.

5.    Move the cursor to an empty cell.  Press the "delete" key.  What do you notice?

Use this result to generate five more confidence intervals using the bootstrap.

Write your intervals below.

# Chapter 10

# Hypothesis Tests Regarding a Parameter

# The Logic of Hypothesis Testing

> Suppose a politician wants to know if a majority (more than 50%) of her constituents are in favor of a certain policy. The politician hires a polling firm to obtain a random sample of 500 registered voters in her district and asks them to disclose whether they are in favor of the policy. What would be convincing evidence that the true (that is, population) proportion of registered voters is greater than 50%?

In this scenario, we are testing

$$H_o: p = 0.5 \text{ versus } H_1: p > 0.5$$

Load the **Political Poll Applet** that is located at www.pearsonhighered.com/sullivanstats to simulate surveying 500 registered voters, assuming that 50% of all registered voters are in favor of the policy and 50% are against the policy. This is done because we always assume that the statement in the null hypothesis is true until we have evidence to the contrary. So we assume that we are sampling from a population where the proportion of registered voters who are in favor of the policy is 0.5.

1.     Explain how 500 coins might be used to simulate the survey results.

2.     Simulate obtaining a simple random sample of 500 voters assuming that $p = 0.5$, exactly one time by clicking "1 run". In this simulation, we are flipping 500 coins where each coin represents a voter. The probability of obtaining a head (a voter in favor of the policy) is 0.5.

       How many heads did you obtain? (Click on the red 1 under "Runs" to see the number of heads.)   That is, how many registered voters are in favor of the policy?

       What is the sample proportion of voters in favor of the policy?

3.     Simulate obtaining a second simple random sample of 500 voters, assuming that $p = 0.5$, exactly one time by clicking "1 run".

       In this sample, how many registered voters are in favor of the policy? (Click on the red 2 under "Runs" to determine the number of heads.)

       What is the sample proportion of voters in favor of the policy?

4.     The sample proportion obtained in part (3) is likely different from the sample proportion obtained in part (2). Why?

5.    Suppose that the polling firm actually conducts a survey of 500 registered voters and finds that 260 are in favor of the policy so that $\hat{p} = \dfrac{260}{500} = 0.52$.

> Do you believe that this sample evidence suggests that the true percentage of registered voters in favor of the policy is greater than 0.5? Or, is it reasonable (that is, not all that unlikely) to obtain a sample of 260 voters in favor of the policy even though the population proportion of voters in favor of the policy is 0.5?

To help answer those questions, simulate conducting a survey of 500 registered voters many times and determine the proportion of times you observe 260 or more in favor of the policy. To do this, "Reset" the applet. Click "1000 runs" to simulate obtaining a simple random sample of 500 registered voters 1000 different times and recording the number of individuals in favor of the policy (assuming that the population proportion is 0.5). In the cell "# of heads," be sure the drop-down menu is set to >=. In the cell to the right of the drop-down menu, type 260 and press Enter. Look in the Count column.

How many of the1000 different surveys resulted in 260 or more voters in favor of the policy?

What proportion of the 1000 different simulations results in a sample proportion of 0.52 or higher even though the population proportion is 0.5?

Do you consider this convincing evidence that the proportion of voters in favor of the policy is greater than 0.5? Why or why not?

What is the shape of the distribution of sample proportions? What is the center? Explain why the shape and center are reasonable.

6.    What if the polling firm discussed in part (5) conducts a survey of 500 registered voters and finds that 270 are in favor of the policy so that $\hat{p} = \dfrac{270}{500} = 0.54$?

Do you believe that this sample evidence suggests that the true proportion of registered voters in favor of the policy is greater than 0.5?

Reset the applet. Click "1000 runs" to simulate obtaining a simple random sample of 500 registered voters 1000 different times and record the number of individuals in favor of the policy (assuming that the population proportion is 0.5). In the cell "# of heads," be sure the drop-down menu is set to >=. In the cell to the right of the drop-down menu, type 270 and press Enter. Look in the Count column.

How many of the 1000 different simulations resulted in 270 or more voters in favor of the policy?

What proportion of surveys results in a sample proportion of 0.54 or higher even though the population proportion is 0.5?

Do you consider this convincing evidence that the proportion of voters in favor of the policy is greater than 0.5? Why or why not?

7.     What if the polling firm discussed in part (5) conducts a survey of 500 registered voters and finds that 280 are in favor of the policy so that $\hat{p} = \dfrac{280}{500} = 0.560$?

Do you believe that this sample evidence suggests that the true proportion of registered voters in favor of the policy is greater than 0.5?

Reset the applet. Click "1000 runs" to simulate obtaining a simple random sample of 500 registered voters 1000 different times and record the number of individuals in favor of the policy (assuming that the population proportion is 0.5). In the cell "# of heads," be sure the drop-down menu is set to >=. In the cell to the right of the drop-down menu, type 280 and press Enter. Look in the Count column.

How many of the 1000 different surveys resulted in 280 or more voters in favor of the policy?

What proportion of surveys results in a sample proportion of 0.56 or higher even though the population proportion is 0.5?

Do you consider this convincing evidence that the proportion of voters in favor of the policy is greater than 0.5? Why or why not?

**Note:** The proportion of simulations that resulted in a sample proportion as extreme, or more extreme, than 0.52 (for simulation from part 5), 0.54 (for simulation from part 6), or 0.56 (for simulation from part 7) is called the **P-value.**

8.     Describe the sampling distribution of $\hat{p}$ under the assumption the statement in the null hypothesis is true.

9.     Compute $P\left(\hat{p} > 0.52\right)$ assuming $H_0$: $p = 0.5$ is true. Compare this result to the simulation from part (5). Repeat this for $\hat{p} = 0.54$ and $\hat{p} = 0.56$. Finally, conclude that either simulation or a normal model (when the requirements are satisfied) may be used to approximate P-values for hypotheses involving a population proportion.

# Interpreting *P*-Values

A *P*-value is the probability of observing a sample statistic as extreme or more extreme than the one observed under the assumption that the null hypothesis is true.  What does this mean in practice though?  Can you interpret and understand a *P*-value in the real world?  This activity will show you one place that you might see *P*-values in the future and will give you the opportunity to try to interpret them.

Find a prescription drug insert either from your local pharmacy or by browsing online.

Find at least two *P*-values in the prescription drug information and cut or print out the section that references the *P*-value.  Attach this here.

Write a few sentences to interpret each *P*-value in the proper context.

# Testing a Claim with Skittles I

> Skittles Brand candies come in red, orange, yellow, green, and purple (violet).  It is well known that M&M Brand candies are not represented with equal frequency.  However, do the colors of Skittles candies occur with equal frequency?

1.  Write the appropriate null and alternative hypotheses to determine if the proportion of each color is the same in a bag of Skittles.

    $H_0:$                          versus          $H_1:$

2.  Randomly choose a color from one of the five colors.  Count the total number of candies in the bag as well as the number for the color you selected.  What is the sample proportion for the color you chose?

    Number of Skittles in the bag:      _____

    Number of Skittles in your color:   _____

    Sample proportion for your color:   _____

3.  Treat the candies in the bag as a simple random sample of all candies produced.  Verify the model requirements to conduct an appropriate hypothesis test.

4.  Test the hypothesis you wrote in step #1.  Be sure to state a conclusion in a complete sentence.

5.  Compare your results to the others in class. Did everyone arrive at the same conclusion for the same color?  What about different colors?

# Understanding Type I Error Rates I

> This activity will utilize an applet to help you understand what it means to make a Type I error and how sample size effects the probability of this type of error.

1.   Load the *Hypothesis Tests for a Proportion (n = 200) Applet* from www.pearsonhighered.com/sullivanstats or from StatCrunch by selecting *Applets > Hypothesis tests > for a proportion*. Let the sample size be $n = 200$, True $p = 0.3$, Null $p = 0.3$, and Alternative set to <. Simulate 1000 simple random samples of size $n = 200$.

   a.   What would it mean to make a Type I error for this hypothesis test?

   b.   Using the $\alpha = 0.05$ level of significance, what proportion of the samples would you expect to lead to a Type I error? What proportion of the samples actually led to a Type I error?

2.   Load the *Hypothesis Tests for a Proportion (n = 400) Applet* from www.pearsonhighered.com/sullivanstats or from StatCrunch by selecting *Applets > Hypothesis tests > for a proportion*. Let the sample size be $n = 400$, True $p = 0.3$, Null $p = 0.3$, and Alternative set to <. Simulate 1000 simple random samples of size $n = 400$.

   a.   Using the $\alpha = 0.05$ level of significance, what proportion of the samples would you expect to lead to a Type I error? What proportion of the samples actually led to a Type I error?

   b.   Compare your results from #1 and #2. Does sample size play a role in the probability of making a Type I error? Explain.

3.   Load the *Hypothesis Tests for a Proportion (n = 50) Applet* from www.pearsonhighered.com/sullivanstats or from StatCrunch by selecting *Applets > Hypothesis tests > for a proportion*. Let the sample size $n = 50$, True $p = 0.1$, Null $p = 0.1$, and Alternative set to >. Simulate 1000 simple random samples of size $n = 50$.

a.    What would it mean to make a Type I error for this hypothesis test?

b.    Using the $\alpha = 0.05$ level of significance, what proportion of the samples would you expect to lead to a Type I error?  What proportion of the samples actually led to a Type I error?

c.    Repeat the simulation for a sample of size $n = 80$.  What proportion of the samples actually led to a Type I error?

d.    Repeat the simulation for a sample of size $n = 120$.  What proportion of the samples actually led to a Type I error?

e.    Explain what happens to the probability of making a Type I error as the sample size increases.

# Testing Cola Preferences

> Do you think Coke or Pepsi is more popular? Or do you think they are preferred in equal numbers? This activity will use a simulation to explore and test the assumption that there is no difference in the proportion of males who prefer Coke versus Pepsi.

1.  In a survey of 20 males conducted at surveycentral.org, 13 indicated they prefer Coke and 7 indicated they prefer Pepsi.

    a.  What proportion of the respondents preferred Coke?

    b.  Assuming there is no difference in the proportion of males who prefer Coke over Pepsi, how many of the 20 males in the survey would we expect to prefer Coke?

2.  Suppose that reality is that there is no difference in the proportion of males that prefer Coke over Pepsi. If this is the case, we would expect half the respondents to choose Coke, and half to choose Pepsi. Is it possible that this is the case, and we just happened to obtain survey results in which 13 out of 20 prefer Coke? Sure, it is possible, but how possible? To get a sense of the likelihood of obtaining 13 out of 20 males that prefer Coke, we will simulate randomly choosing 20 males from a population in which half prefer Coke and half prefer Pepsi. Essentially, this would be like flipping a coin 20 times and recording the number of heads (for Coke) and the number of tails (for Pepsi).

    a.  Flip a coin 20 times and record the number of heads (the number of Coke drinkers) and tails (number of Pepsi drinkers).

    | Number of heads (Number who prefer Coke) | Number of tails (Number who prefer Pepsi) |
    | --- | --- |
    |  |  |

    b.  Repeat this experiment two more times so that you have three trials with 20 coin flips each.

    | Number of heads (Number who prefer Coke) | Number of tails (Number who prefer Pepsi) |
    | --- | --- |
    |  |  |
    |  |  |

    c.         Compile the data from your entire class. Draw a dot plot that shows the number of heads (number who prefer Coke) for each trial conducted.

    d.         What proportion of the trials resulted in 13 or more heads (13 or more individuals who prefer Coke)?

3.        To get a true sense as to the likelihood of 13 out of 20 males choosing Coke even if there really is no cola preference, we need to simulate the result thousands of times. Rather than using a coin to perform this simulation, we can use an applet.

    a.         Open the *Cola Applet* at www.pearsonhighered.com/sullivanstats or from StatCrunch by selecting *Applets > Simulation > Coin Flipping Applet*. Set probability of heads to 0.5, number of coins to 20, and number to $\geq 13$. Hit Compute! And then select 1000 runs.

    b.         Press the "simulate" button and determine the proportion of repetitions for which we observed 13 or more heads. The result of your simulation represents an approximate *P*-value for the hypothesis test.

    c.         What is the result you obtained? What does this result imply? Does this result suggest that more males prefer Coke over Pepsi?

4.        Suppose the survey resulted in 16 out of 20 males choosing Coke over Pepsi. How would your conclusions change?

5.        Use the binomial probability formula to compute the probability of obtaining 13 or more successes in 20 trials of an experiment for which the probability of success is 0.5. How does this result compare with the simulation results?

# Testing a Claim with Beads

> One purpose of this activity is to make connections between the *P*-Value and confidence interval methods for testing a claim about a proportion. Also, it will give you an opportunity to better understand Type I errors.

1.  Use a small Dixie cup and obtain a sample of beads from the bucket.   Treat this as a simple random sample. Count the number of blue beads and the total number of beads in your cup.   Then determine the proportion of your beads that are blue.

    Number of blue beads: _____
    Total number of beads: _____
    Proportion of blue beads (round to three decimal places): _____

2.  a.  Suppose you want to test the hypotheses $H_o : p = p_o$ versus $H_1 : p \neq p_o$ where $p_o$ is the population proportion of blue beads in the bucket.   Verify that the requirements for using the normal model to determine a *P*-value for this hypothesis test are satisfied.   Ask your instructor to disclose the number of beads in the bucket.

    b.  Draw a normal model that describes the sampling distribution of $\hat{p}$ .  By hand, use this model to approximate the *P*-value of the hypothesis test from part c.

    c.  Based on the *P*-value from part b, state a conclusion to your hypothesis test.

    d.  Did you make a Type I error?  How do you know this?

    e.  Compare your results with the rest of the class.  What proportion of groups made a Type I error?

3.    a.    Using the data from question 1, test the hypotheses $H_o : p = p_o$ versus $H_1 : p \neq p_o$ for the range of potential values of $p_0$ supplied by your instructor. For which values of $p_0$ do you reject the null hypothesis at the 0.05 level of significance?

      b.    Construct a 95% confidence interval for the proportion of blue beads in the bucket. Include an interpretation of the interval.

      c.    Compare the lower bound and upper bound for the confidence interval in part b to the values of $p_0$ for which the null hypothesis is rejected. What do you notice?

4.    Use the data from question 1 to test the hypothesis $H_o : p = p_o$ versus $H_1 : p \neq p_o$ for a value of $p_0$ supplied by your instructor.

      Did you reject the null hypothesis for this value of $p_0$?

      Use this result to explain the difference between the statements "Accept the null hypothesis" and "Do not reject the null hypothesis."

*Courtesy of Mary Brown, Associate Professor of Mathematics, HACC, Central Pennsylvania's Community College

# Analyzing a Research Article I

To investigate the role of e-cigarettes in smoking cessation, research has been done to "investigate whether the use of e-cigarettes is associated with smoking cessation or reduction, and whether there is any difference in efficacy of e-cigarettes with and without nicotine on smoking cessation."

The results of a meta-analysis showed that e-cigarettes with nicotine were more effective than those without nicotine for smoking cessation and that the use of e-cigarettes with nicotine was positively associated with smoking cessation and a reduction in the number of cigarettes.

Source: Rahman MA, Hann N, Wilson A, Mnatzaganian G, Worrall-Carter L (2015) E-Cigarettes and Smoking Cessation: Evidence from a Systematic Review and Meta-Analysis. PLoS ONE 10(3): e0122544. doi:10.1371/journal.pone.0122544

Read the full article here: http://journals.plos.org/plosone/article?id=10.1371/journal.pone.0122544

1.   The following table summarizes some of the results in the meta-analysis for smokers who used e-cigarettes with nicotine.

|  | Total Sample | Number Who Stopped Smoking | Proportion (95% CIs) |
|---|---|---|---|
| Study 1 | 464 | 93 | (0.16, 0.24) |
| Study 2 | 289 | 21 | (0.04, 1.10) |
| Study 3 | 200 | 22 | (0.07, 0.15) |
| Study 4 | 35 | 16 | (0.29, 0.62) |
| Study 5 | 40 | 5 | (0.02, 0.23) |
| Study 6 | 214 | 67 | (0.31, 0.38) |

a.   Use technology to confirm the 95% confidence intervals shown in the table for the proportion of smokers who stopped smoking in each study.

b.   Find the total number of smokers in all six studies and the total number who stopped smoking. Find a new confidence interval for the proportion of smokers who stopped smoking using these results.

Total number of smokers:

Total number who quit:

Overall proportion of smokers who quit:

Confidence interval for proportion who quit:

Confirm your proportion with the information given in the full article.

2.    Use the combined results to test the claim that fewer than 20% of smokers who use e-cigarettes with nicotine are able to stop smoking.

# Testing a Claim with Skittles II

Skittles Brand candies can be purchased in 1-pound bags. But do you really get exactly one pound of Skittles? This activity will help you test that claim from the manufacturer.

1.    Write the appropriate null and alternative hypotheses to determine if the mean weight of a bag of Skittles is one pound.

   $H_0$ :                              versus              $H_1$ :

2.    Weigh your bag of Skittles as accurately as possible.

   Exact weight of your bag of Skittles:          _____

3.    Treat all the bags of Skittles in the class as a simple random sample of all such bags produced. Verify the model requirements to conduct an appropriate hypothesis test.

4.    Record the weight of each bag in the class.

5.    Test the hypothesis you wrote in step #1. Be sure to state a conclusion in a complete sentence.

# Understanding Type I Error Rates II

> This activity will utilize an applet to help you understand what it means to make a Type I error when a hypothesis is conducted on a mean. You will explore how sample size and violating assumptions effect the probability of this type of error.

1.   Load the *Hypothesis Tests for a Mean (n = 10) Applet* from www.pearsonhighered.com/sullivanstats or from StatCrunch by selecting *Applets > Hypothesis Tests > for a mean*. Set the shape to normal, the mean to 100, and the standard deviation to 15. These parameters describe the distribution of IQ scores.

    a.   Obtain 1000 simple random samples of size $n = 10$ from this population, and test whether the mean is different from 100.

    b.   How many of the samples would you expect to lead to a rejection of the null hypothesis if the level of significance is $\alpha = 0.05$?

    c.   How many of the 1000 samples led to rejection of the null hypothesis?

2.   Load the *Hypothesis Tests for a Mean (n = 30) Applet* from www.pearsonhighered.com/sullivanstats or from StatCrunch by selecting *Applets > Hypothesis Tests > for a mean*. Set the shape to normal, the mean to 100, and the standard deviation to 15. These parameters describe the distribution of IQ scores.

    a.   Obtain 1000 simple random samples of size $n = 30$ from this population, and test whether the mean is different from 100.

    b.   How many of the samples would you expect to lead to a rejection of the null hypothesis if the level of significance is $\alpha = 0.05$?

    c.   How many of the 1000 samples led to rejection of the null hypothesis?

3.   Compare the results of #1 and #2. Does the sample size have any effect on the number of samples that incorrectly rejected the null hypothesis?

4.   Load the *Hypothesis Tests for a Mean (Skewed with n = 6) Applet* from www.pearsonhighered.com/sullivanstats or from StatCrunch by selecting *Applets > Hypothesis Tests > for a mean*. Set the shape of the distribution to right skewed, the mean to 50, and the standard deviation to 10.

    a.          Obtain 3000 simple random samples of size $n = 6$ from this population. Test whether the mean is different from 50.

    b.          How many of the samples led to rejection of the null hypothesis using a level of significance of $\alpha = 0.05$? How many would you expect to lead to rejection of the null hypothesis?

5.        Load the *Hypothesis Tests for a Mean (Skewed with n = 60) Applet* from www.pearsonhighered.com/sullivanstats or from StatCrunch by selecting *Applets > Hypothesis Tests > for a mean*. Set the shape of the distribution to right skewed, the mean to 50, and the standard deviation to 10.

    a.          Obtain 1000 simple random samples of size $n = 60$ from this population. Test whether the mean is different from 50.

    b.          How many of the samples led to rejection of the null hypothesis using a level of significance of $\alpha = 0.05$? How many would you expect to lead to rejection of the null hypothesis?

6.        Explain why the results of #4 and #5 differ.

# Using Bootstrapping to Test a Claim

In Chapter 9, you learned how to use bootstrapping to construct a confidence interval about a parameter. We can also use bootstrapping to test a hypothesis about a parameter. In this activity, you will use bootstrapping to test a hypothesis about a population mean.

The Environmental Protection Agency (EPA) reports that the combined (highway and city driving) miles per gallon (mpg) of a 2010 Ford Fusion Hybrid with a 2.5 liter engine and automatic transmission is 39 mpg. The website fueleconomy.gov allows individuals to report the combined miles per gallon that they are obtaining. The following data represents the miles per gallon of 15 individuals. We want to know whether the data suggest the combined miles per gallon is less than the suggested mpg of 39.

| 34.1 | 36.7 | 36.7 | 34.9 | 30.3 |
|------|------|------|------|------|
| 37.2 | 37.2 | 40.0 | 36.7 | 41.0 |
| 35.0 | 39.7 | 35.6 | 42.0 | 39.8 |

Source: www.fueleconomy.gov

The hypotheses we are testing are

$$H_o : \mu = 39 \text{ versus } H_1 : \mu < 39$$

The sample mean of the data is $\bar{x} = 37.13$ mpg and the sample standard deviation is $s = 3.03$ mpg.

We compute the test statistic as

$$t_0 = \frac{37.13 - 39}{3.03 / \sqrt{15}} = -2.390$$

The challenge in using the bootstrap to test the hypothesis is that the procedure requires us to assume the data come from a population whose mean is stated in the null hypothesis. Therefore, we need to adjust our data so that it has a mean of 39. This can be accomplished by subtracting the sample mean from the data and then adding the mean stated in the null. For our data, this would mean computing

$$\tilde{x}_i = MPG_i - \overline{MPG} + 39$$

for $i = 1, 2, ..., 15$. The transformed data is shown in the table below.

| 35.97 | 38.57 | 38.57 | 36.77 | 32.17 |
|-------|-------|-------|-------|-------|
| 39.07 | 39.07 | 41.87 | 38.57 | 42.87 |
| 36.87 | 41.57 | 37.47 | 43.87 | 41.67 |

If we compute the mean of the data in the table above, we obtain a value of 39 mpg, which is what we want. So, we now obtain 1000 bootstrap samples from this data. The first three bootstrap samples are shown in the table below.

| 37.550 | 38.543 | 40.043 | 38.803 | 39.097 | 39.683 | 39.550 | 39.437 | 39.757 | 38.330 | 39.117 | 39.163 | 37.470 | 39.403 | 38.757 |
|--------|--------|--------|--------|--------|--------|--------|--------|--------|--------|--------|--------|--------|--------|--------|
| 40.010 | 38.577 | 38.777 | 38.457 | 39.330 | 39.963 | 38.837 | 39.290 | 39.077 | 39.103 | 38.710 | 38.857 | 37.330 | 39.117 | 37.423 |
| 38.143 | 39.097 | 39.110 | 38.503 | 40.763 | 39.503 | 39.590 | 39.163 | 39.297 | 39.243 | 38.650 | 39.670 | 39.757 | 39.623 | 40.063 |

We compute the sample mean, $\bar{x}$, and sample standard deviation, $s_{\bar{x}}$, which are used to obtain the test statistic,

$$t_o = \frac{\bar{x} - 39}{\frac{s_{\bar{x}}}{\sqrt{15}}}$$, for each bootstrap sample. After arranging the test statistics in ascending order, we determine the

proportion that are less than $t_o = \frac{37.13 - 39}{\frac{3.03}{\sqrt{15}}} = -2.390$, which represents the approximation of the $P$-value.

From the MINITAB output below, we see that 13 of the bootstrap samples result in a test statistic less than -2.390. Our approximate $P$-value is $13/1000 = 0.013$.

| | C1 | C2 | C3 | C4 | C5 | C6 | C7 | C8 |
|---|---|---|---|---|---|---|---|---|
| ↓ | Transformed MPG | Bootstrap | Sample | ByVar1 | Mean1 | StDev1 | t | ordered t |
| 1 | 35.97 | 38.57 | 1 | 1 | 39.6700 | 2.21392 | 1.17208 | -5.23784 |
| 2 | 39.07 | 41.57 | 1 | 2 | 38.6567 | 3.56127 | -0.37338 | -4.92661 |
| 3 | 36.87 | 38.57 | 1 | 3 | 38.3367 | 3.86387 | -0.66490 | -3.91153 |
| 4 | 38.57 | 41.67 | 1 | 4 | 38.6433 | 2.73926 | -0.50428 | -3.63501 |
| 5 | 39.07 | 42.87 | 1 | 5 | 39.8367 | 2.30517 | 1.40571 | -3.16964 |
| 6 | 41.57 | 42.87 | 1 | 6 | 39.7300 | 3.73436 | 0.75710 | -2.83561 |
| 7 | 38.57 | 39.07 | 1 | 7 | 39.9967 | 2.30913 | 1.67165 | -2.74463 |
| 8 | 41.87 | 41.87 | 1 | 8 | 37.8700 | 2.95876 | -1.47916 | -2.49048 |
| 9 | 37.47 | 36.77 | 1 | 9 | 39.4700 | 2.00036 | 0.90999 | -2.47675 |
| 10 | 36.77 | 37.47 | 1 | 10 | 39.7433 | 2.26352 | 1.27188 | -2.46000 |
| 11 | 38.57 | 41.87 | 1 | 11 | 39.0033 | 3.36254 | 0.00384 | -2.43356 |
| 12 | 43.87 | 38.57 | 1 | 12 | 39.1767 | 3.41373 | 0.20043 | -2.41589 |
| 13 | 32.17 | 39.07 | 1 | 13 | 38.8900 | 2.49434 | -0.17080 | -2.39437 |
| 14 | 42.87 | 36.77 | 1 | 14 | 38.9700 | 2.09523 | -0.05545 | -2.26424 |
| 15 | 41.67 | 37.47 | 1 | 15 | 38.8300 | 3.20509 | -0.20543 | -2.25058 |
| 16 | | 38.57 | 2 | 16 | 39.3567 | 3.28478 | 0.42053 | -2.19008 |
| 17 | | 35.97 | 2 | 17 | 39.2367 | 2.91172 | 0.31480 | -2.18749 |
| 18 | | 38.57 | 2 | 18 | 37.7633 | 3.28557 | -1.45777 | -2.16002 |

Using a one-sample $t$-test to test $H_o : \mu = 39$ versus $H_1 : \mu < 39$, we find the $P$-value to be 0.016.

Now let's have you try one. The following data represents the miles per gallon that 10 individuals experienced with their 2009 Smart car with a 1.0 Liter engine and automatic transmission.   We want to know whether the data suggest the combined miles per gallon is greater than the suggested mpg of 36.

| 34.0 | 34.6 | 39.8 | 36.6 | 42.9 |
| 41.5 | 46.3 | 35.3 | 32.3 | 43.8 |

Source: fueleconomy.gov

1.    Determine the appropriate null and alternative hypotheses.

2.    Compute the test statistic.

3.    Transform the original data so that it has a mean equivalent to the mean stated in the null hypothesis.

|  |  |  |  |
|  |  |  |  |

4.    Obtain 1000 bootstrap samples using a statistical spreadsheet.

5.    Determine the mean and standard deviation of each bootstrap sample.

6.    Compute the test statistic for each bootstrap sample.

7.    After arranging the test statistics in ascending order, determine the approximate P-value by finding the proportion of test statistics greater than the value of the test statistic found in #2.

8.    Find the P-value using the single sample t-test.

9.        Explain why the original data must be transformed.

10.       Devise an approach to use for using bootstrapping on a two-sided hypothesis test.

# Computing the Power of a Test

This applet activity will help you explore how the power of a hypothesis test changes as the population proportion approaches the proportion stated in the null hypothesis.

Load the **Hypothesis Tess for a Proportion (p = 0.44) Applet** from www.pearsonhighered.com/sullivanstats or from StatCrunch by selecting **Applets > Hypothesis Tests > for a proportion**. Set the sample size to $n = 100$, True $p$ to 0.44, Null $p$ to 0.5, and alternative to <.

a.     Simulate obtaining 2000 simple random samples of size $n = 100$.
        What proportion of samples led to correctly rejecting the null hypothesis?
        That is, what is the power of the test?

b.     Compute the power of the test for the simulation presented in part (a) by hand.

c.     Repeat part (a) for True $p$ of 0.46 and 0.48.

d.     What happens to the power of the test as the true value of the population proportion
        approaches the proportion stated in the null hypothesis?

# Chapter 11

# Inferences on Two Samples

# Making an Inference about Two Proportions

A study was conducted by researchers to determine if application of duct tape is more effective than cryotherapy (liquid nitrogen applied to the wart for 10 seconds every 2 to 3 weeks) in the treatment of common warts.

The researchers randomly divided 40 patients into two groups. The 20 patients in Group 1 had their warts treated by applying duct tape to the wart for 6.5 days and then removing the tape for 12 hours, at which point the cycle was repeated, for a maximum of 2 months. The 20 patients in Group 2 had their warts treated by cryotherapy for a maximum of six treatments.

Once the treatments were complete, it was determined that 17 patients in Group 1 (duct tape) and 12 patients in Group 2 (cryotherapy) had complete resolution of their warts. *Source*: Dean R. Focht III, Carole Spicer, Mary P. Fairchok. "The Efficacy of Duct Tape vs. Cryotherapy in the Treatment of Verruca Vulgaris (the Common Wart)," *Archives of Pediatrics and Adolescent Medicine,* 156(10), 2002.

1.  What are the researchers trying to show in this study?

2.  Explain why the hypotheses to be tested are $H_0$: $p_1 = p_2$ versus $H_1$: $p_1 > p_2$.

3.  What proportion of the subjects in each group experienced complete resolution of their warts? What is the difference in sample proportions, $\hat{p}_{duct} - \hat{p}_{cryotherapy}$?

4.  What does the statement in the null hypothesis suggest about the effectiveness of the two treatments?

5.  We need to determine whether the observed differences are due to random error (and there really is no difference in the treatments), or if the differences are significant (so that one treatment is superior to the other). To answer this question, we will randomly assign a treatment to each of the outcomes. The idea is that under the assumption the null hypothesis is true, it should not matter whether a success is the result of duct tape or cryotherapy.

    a.  Take a standard deck of cards, and let 20 red cards represent the duct tape, and let 20 of the black cards represent cryotherapy. Shuffle the cards and then deal 29 cards. These 29 cards represent the 29 individuals who had complete resolution of their warts.

    b.  Record the proportion of red cards (duct tape) that had complete resolution of their warts. Record the proportion of black cards (cryotherapy) that had complete resolution of their warts. Now, determine the difference in the proportions, $\hat{p}_{duct} - \hat{p}_{cryotherapy}$.

        Is this proportion as extreme (or more extreme) than the observed difference in proportions?

6.    a.    Combine your results from the card simulation with the rest of the class. What proportion of simulations resulted in a sample difference as extreme or more extreme than those observed?

      b.    Would you be willing to conclude that duct tape is superior to cryotherapy with this number of observations?

7.    What we really want to know is, "If we conducted this experiment many, many times, how likely is it to obtain a sample difference in proportions as extreme, or more extreme than the proportion I observed if there is no difference in the treatments." Certainly, doing this physical simulation many, many times would be labor intensive. So, we will rely on an applet to conduct the simulation for us.

      a.    Open the *Randomization Test Warts Applet* from www.pearsonhighered.com/sullivanstats. To create the applet in StatCrunch, first enter the raw data by creating a column titled "Treatment." Create 20 entries called "Duct" and 20 entries called "Cryotherapy." Create a second column called "Outcome." Enter 17 Successes and 3 Failures for Duct; 12 Successes and 8 Failures for Cryotherapy. Select *Applets > Resampling > Randomization test for two proportions*. Select the "From data table" radio button. Under "Sample 1 in" choose "Outcome" from the drop-down menu and type "Treatment = Duct" under "Where." Under "Sample 2 in," choose "Outcome" from the drop-down menu and type "Treatment = Cryotherapy" under "Where." Type "Success" in the Success cell. Click Compute!

      b.    Press randomize 1 time. What is the difference in the sample proportions?

      c.    Press randomize 1 time again. Did you get a different difference in the sample proportions?

      d.    Now press randomize 1000 times. The applet draws a histogram of the differences in proportions and records the number of simulations that result in differences in the sample proportions as extreme, or more extreme, than the sample difference found in #3.

            Based on the simulation, what might you conclude?

            Are you willing to conclude duct tape is a more effective treatment in dealing with the removal of warts? Explain your position.

# Analyzing Rates of Drug Side Effects

Have you ever read the information insert that comes with your prescription medication? Have you ever wondered if the side effects listed are really more likely to occur with the medication than without the medication? This activity will help you analyze the side effect rate for a drug and determine if there is a statistically significant difference in the side effect rates for the drug and placebo groups.

1. Suppose we want to determine which adverse reactions were more common with the prescription drug sildenafil citrate than the placebo, using a significance level of 0.05.

| Adverse Event | Percentage of Patients Reporting Event | |
|---|---|---|
| | Sildenafil citrate N=734 | Placebo N=725 |
| Headache | 16% | 4% |
| Flushing | 10% | 1% |
| Dyspepsia | 7% | 2% |
| Nasal Congestion | 4% | 2% |
| Urinary Tract Infection | 3% | 2% |
| Abnormal Vision | 3% | 0% |
| Diarrhea | 3% | 1% |
| Dizziness | 2% | 1% |
| Rash | 2% | 1% |

a. For which side effect are the conditions for the hypothesis testing procedure **not** met? Explain. Do not do the test on that side effect and leave that line in the table blank.

b. For each side effect listed in the table, test if the proportion of the sildenafil citrate users who experienced the side effect is higher than the proportion of the placebo group who experienced the side effect.

Summarize your results in the table below.

| Adverse Event | P-value | Significantly higher rate for sildenafil citrate users? (Yes/No) |
|---|---|---|
| Headache | | |
| Flushing | | |
| Dyspepsia | | |
| Nasal Congestion | | |
| Urinary Tract Infection | | |
| Abnormal Vision | | |
| Diarrhea | | |
| Dizziness | | |
| Rash | | |

2.    Do the same treatment for a different prescription drug of your choice. Write up one test completely and summarize the others in the table at the bottom of the page. You can find this information online or you can ask at a local pharmacy for extra prescription information inserts. Either way you obtain the data, please include the information here, in some kind of table or as a screen capture.

| Adverse Event | *P*-value | Significantly higher rate for prescription drug users? (Yes/No) |
|---|---|---|
|  |  |  |
|  |  |  |
|  |  |  |
|  |  |  |
|  |  |  |
|  |  |  |
|  |  |  |

# Predicting Basketball Game Outcomes

> Two famous college basketball analysts are Jay Bilas and Dick Vitale. Both tend to make predictions regarding the outcome of college basketball games with varying degrees of success. In 116 games, Bilas correctly predicted 77 of them; in 116 games, Vitale correctly predicted 69 of them.
>
> Does this mean Bilas is better at predicting games than Vitale?

1. Suppose we would like to know whether the evidence suggests if Bilas is better at predicting the outcome of college basketball games. What would be the null and alternative hypothesis?

2. What proportion of correct predictions were made by Bilas? What proportion of correct predictions were made by Vitale? What is the value of the statistic $\hat{p}_{Bilas} - \hat{p}_{Vitale}$? Explain why this is a statistic and identify some sources of variability in the statistic.

3. Why is this an independent sample? Explain why a dependent sample would be a better approach.

4. Rather than using the normal model to estimate the P-value, we can use a randomization applet. Because the null hypothesis is a statement of no difference, the outcome of any given prediction could just as easily come from Bilas as Vitale because the null hypothesis states there is no difference in the proportion of correct predictions.

   a. Go to www.pearsonhighered.com/sullivanstats and open the applet *Randomization Test Basketball*. Or, go to www.pearsonhighered.com/sullivanstats to obtain the data file "Basketball." Open the file in StatCrunch. Then, in StatCrunch, select *Applets > Resampling > Randomization test for two proportions*.

   Fill in the window that appears as shown and click Compute!.

b.     Click "1 time" and observe the window. The applet randomly assigns the analyst (Bilas or Vitale) to an outcome (Correct or Wrong).

In the random assignment, what proportion of the Correct predictions went to Bilas? to Vitale?

What is the value of $\hat{p}_{Bilas} - \hat{p}_{Vitale}$ ?

c.     Explain how this random assignment of outcomes is consistent with the statement in the null hypothesis.

d.     Click "1 time" and observe the window. In the random assignment, what proportion of the Correct predictions went to Bilas? to Vitale?

What is the value of $\hat{p}_{Bilas} - \hat{p}_{Vitale}$ ? The results likely differ from those in part (b). Why?

e.     We want to obtain a measure of the likelihood of obtaining a sample statistic as extreme or more extreme than the one observed under the assumption the two individuals have the same success rate. We do this by randomly assigning outcomes many, many times. Click "1000 times".

What proportion of the randomizations above the difference in sample proportions found in part (b)?

What does this value represent?

5.     a.     What is the shape of the difference in sample proportions?

Verify that a normal model could be used to approximate the P-value for the hypothesis test from part (a).

What is the assumed center and spread of the difference in sample proportions that is used to approximate the P-value?

b.     Use the normal model to approximate the P-value for the hypothesis test from part (a). Interpret the P-value.

6.    Based on your analysis, does the sample evidence suggest that Bilas is better than Vitale at making predictions?

7.    Why do the approximate *P*-values from parts (4e) and (5b) differ?

# Considering the Effects of Grammar

See Problem 31 from section 11.1 in the text:

Psychologists Teenie Matlock and Caitlin Fausey asked students to read two sentences about hypothetical politicians. Ninety-eight students read, "Last year, Mark was having an affair with his assistance and was taking hush money from a prominent constituent." Let's call this sentence A. ninety-four other students read, "Last year, Mark had an affair with his assistant and took hush money from a prominent constituent." We will call this sentence B. In the study, 71 of the 98 students who read sentence A and 49 of the 98 students who read sentence B felt the politician would not be re-elected. Source: Fausey, C.M., and Matlock, T. Can Grammar Win Elections? Political Psychology, no. doi: 10.1111/j. 1467-9221.2010.00802.x.

In this activity, you will use a randomization applet to investigate whether or not the results suggest that the sentence structure makes a difference in deciding whether the politician would be re-elected.

Go to www.pearsonhighered.com/sullivanstats and open the applet *Randomization Test Sentence*. Or, in StatCrunch, select *Applets > Resampling > Randomization test for two proportions*. Select the "From summary" radio button. Let Sentence A represent sample 1. Enter 71 for the number of successes and 98 for the number of observations. Let Sentence B represent sample 2. Enter 49 for the number of successes and 94 for the number of observations. Click Compute!.

1.    What is the difference in sample proportions, $\hat{p}_1 - \hat{p}_2$?

2.    If the null hypothesis is true, what would we expect the sample difference in proportions to equal?

3.    Click "1 time." The applet randomly assigns either Sample 1 or Sample 2 to an outcome. How is this consistent with the assumption stated in the null hypothesis? In your randomization, what was the sample difference in proportions?

4.    Click "1000 times." What proportion of randomizations were "at or below" or "at or above" the observed difference in sample proportions found in part (1)? Interpret this result.

# Analyzing Kissing Data

> Psychologists at the State University of New York found that 59% of men and 66% of women have ended a budding relationship because a kiss did not go well.  The author of your text found this result difficult to believe, so he administered a survey to a random sample of adult Americans and asked each individual to disclose whether they ended a budding relationship because a kiss did not go well.

Go to www.pearsonhighered.com/sullivanstats to obtain the data file "Kiss."

1.      What proportion of males in the survey ended a budding relationship because a kiss did not go well? What proportion of females?

2.      Clearly, the results of the author survey are not consistent with those of the psychologists at the State University of New York.  However, we may still determine if there is a difference between males and females when it comes to this question.  Explain why the normal model may not be used to test whether there is a difference in the proportion of males and females who would end a budding relationship because a kiss did not go well.

3.      Go to www.pearsonhighered.com/sullivanstats and open the applet *Randomization Test Kiss*.  Or, in StatCrunch, select *Applets* > *Resampling* > *Randomization test for two proportions*.  Select the "From data table" radio button.  Fill in the dialogue box as shown and click Compute!.

What is the difference in the sample proportions?

Verify this difference.

4.  If we want to determine if there is a difference in the proportion of males and females who would end a budding relationship because a kiss did not go well, what would be the null and alternative hypotheses?

5.  Run the applet 1 time. Explain what the applet is doing.

6.  Run the applet 5000 times by clicking "1000 times" five times.

    What proportion of the randomizations result in a sample difference as extreme or more extreme than the sample difference in (3)?

    What does the proportion suggest?

7.  Suppose the survey was conducted in which 24 of 105 males and 63 of 177 females indicated they would end a budding relationship because a kiss did not go well. Verify these sample proportions are the sample as those found in part (1).

8.  Use the "From summary" feature in the randomization test for two proportions applet in StatCrunch to randomize 5000 trials of this survey.

    What is proportion of the randomizations resulted in a sample difference as extreme or more extreme than 0.1274?

    Explain what this result suggests.

9.  Verify that a normal model may be used to test the hypotheses stated in (4). Use the normal model to approximate the $P$-value. Explain what this result suggests.

10. Explain the role sample size plays in the ability to reject a statement in the null hypothesis.

# Analyzing a Research Article II

A study was conducted to investigate the effects of psychological stress (general stress, stress at home, and stress at work) on hypertension for middle-aged Chinese men and women.

The results showed that psychological stress is associated with a higher risk of hypertension although not necessarily significantly for each type for each gender.

Source: Hu B, Liu X, Yin S, Fan H, Feng F, Yuan J (2015) Effects of Psychological Stress on Hypertension in Middle-Aged Chinese: A Cross-Sectional Study. PLoS ONE 10(6): e0129163. doi:10.1371/journal.pone.0129163

Read the full article here: http://journals.plos.org/plosone/article?id=10.1371/journal.pone.0129163

1.   The article states that there were 914 participants with hypertension. Of the women, 410 (17.38%) had hypertension, and 504 (13.93%) of the men did as well.

Is the proportion of women with hypertension significantly higher than the proportion of men with hypertension?

2.   Look at the results shown in Table 2.

   a.   Was the hypertension rate significantly higher for people who reported "general stress" compared to those who didn't? What about for individual genders? Confirm the $p$-values shown in the table and interpret them.

b.    Was the hypertension rate significantly higher for people who reported "stress at work" compared to those who didn't? What about for individual genders? Confirm the $p$-values shown in the table and interpret them.

c.    Was the hypertension rate significantly higher for people who reported "stress at home" compared to those who didn't?  What about for individual genders? Confirm the $p$-values shown in the table and interpret them.

3.    In the "Strengths and Limitations" section of the article, it is mentioned that no causal relationships can be inferred from this research since the study had a cross-sectional design. Explain what this means.

# Comparing Arm Span and Height

Most people have an arm span approximately equal to their height. This activity will use your class data to test the claim that the mean arm span is the same as the mean height.

1.    Each student in your class should measure the length of their arm span and their height, both as accurately as possible in centimeters. Collect the paired data (arm span, height) and record it in the first two columns. Then calculate the differences in arm span and height and record them in the third column.

| Arm span (centimeters) | Height (centimeters) | Difference, $d$ (arm span – height) |
|---|---|---|
| | | |
| | | |
| | | |
| | | |
| | | |
| | | |
| | | |
| | | |
| | | |
| | | |
| | | |
| | | |
| | | |
| | | |
| | | |
| | | |
| | | |
| | | |
| | | |
| | | |
| | | |
| | | |
| | | |
| | | |
| | | |
| | | |

2.    Write the hypotheses necessary to test if there is a difference in the mean arm span and the mean height.  Also state the direction of the test.

3.    Conduct the hypothesis test at a significance level of 0.05.  State the *P*-value.

4.    State your conclusion in a complete sentence.

# Using Randomization Test for Independent Means

> Do you think male or female students spend more time on homework? Or do you think there's no difference? How would we test this statistically? This activity will guide you through a randomization test for independent means to analyze this issue.

Professor Sullivan noticed in his Elementary Algebra course that the female students appeared to work longer and more diligently on homework assignments than the male students. To confirm his suspicion, he randomly selected 12 male and 12 female student records in MyMathLab and recorded the amount of time on task for a particular section. The results, in minutes, are below.

| MALE | | | |
|----|-----|----|----|
| 81 | 104 | 60 | 83 |
| 79 | 88  | 88 | 78 |
| 81 | 74  | 72 | 79 |

| FEMALE | | | |
|-----|-----|-----|----|
| 92  | 77  | 107 | 99 |
| 123 | 152 | 116 | 75 |
| 106 | 81  | 70  | 99 |

*Source*: MyMathLab

1.  Draw dot plots of each data set using the same scale. Does it appear that males are spending less time on their homework? Explain.

2.  Calculate the mean time spent on homework for each gender. Then compute the sample mean difference as "Male minus Female." Based on the means, does it appear that males spend less time on their homework? Explain.

3.  Certainly, the dotplot and means suggest that males are not spending as much time on their homework. But, are we willing to judge that, in general, males spend less time on homework? Or, is it possible that each gender spends the same amount of time on homework, and we just happened to select some "less than diligent" males? In other words, are the differences between the two samples statistically significant?

    a.  Two help answer this question, let's do the following exercise. Take 24 index cards and write the time for each individual on a single card. Shuffle the 24 cards and distribute 12 to the "male" group and 12 to the "female" group. Determine the mean for the "male" group. Determine the mean for the "female" group. Finally, compute the difference in the means as "Male minus Female."

    b.  Collect the mean difference from the rest of your classmates. Draw a dotplot of the mean differences below.

c.    Does it appear to be the case that the differences are centered around 0? Why does this make sense?

d.    Are any of the results as extreme, or more extreme, than the sample mean difference obtained in #2? What does this suggest?

4.    Go to www.pearsonhighered.com/sullivanstats and open the applet *Randomization Test Algebra*. Or, in StatCrunch, enter the data above. Select *Applets* > *Resampling* > *Randomization test for two means*. Enter Males for Sample 1; enter Females for Sample 2. Notice the applet determines the mean difference. Verify the sample means and mean difference from #2.

a.    Click Randomize 1 time. Explain what the applet is doing when the randomize button is selected. What is the mean difference?

b.    Now select Randomize 1000 times. What is the center of the histogram of the 1000 (and one) randomized mean differences? Is this what you would expect? What is the shape of the histogram? Is this what you would expect?

c.    How many of your simulated differences result in a mean difference at or below the value found in #2? What proportion of the randomized differences result in a mean difference at or below the value found in #2?

d.    Find the *P*-value for the test using Student's *t*-distribution. How does this result compare to the result from part (c)?

5.    Write a conclusion about the study time put in by males versus females.

# Comparing Bull and Bear Markets

The stock market can sometimes go through periods known as *bull markets* or *bear markets*. A **bull market** exists when the stock market increases in value by at least 20%, while a **bear market** exists when the stock market decreases in value by at least 20%.

The data below represent the number of months a random sample of bull and bear markets have lasted.

**Bear Months**

| 1.9 | 8.3 | 3.3 | 3.4 | 6.8 |
|-----|-----|------|------|------|
| 5.8 | 3.1 | 13.4 | 12.9 | 5.1 |
| 7.6 | 17.9 | 11.8 | 12.1 | 14.9 |
| 6.5 | 8.0 | 18.1 | 21.1 | 20.7 |
| 3.4 | 18.2 | 9.3 | 13.6 | 2.1 |

**Bull Months**

| 4.9 | 2.3 | 0.8 | 1.2 | 2.3 |
|------|------|------|------|------|
| 4.7 | 3.7 | 24.2 | 7.4 | 6.6 |
| 5.0 | 49.7 | 13.1 | 86.9 | 50.4 |
| 44.1 | 26.1 | 32.0 | 74.9 | 61.3 |
| 149.8 | 3.5 | 60.9 | 1.6 | 48.6 |

1. Suppose we would like to know whether bull markets tend to last longer than bear markets. What would be the null and alternative hypotheses?

2. What is the mean length of time a bear market lasts according to the sample data? What is the mean length of time a bull market lasts according to the sample data? What is the difference in sample means, $\bar{x}_{Bull} - \bar{x}_{Bear}$?

3. Why is this an independent sample? Explain why a dependent sample could not be used to answer the research objective posed in question 1.

4. Draw a side-by-side boxplot of the number of months each market lasts. Does there appear to be a difference in the median time each market lasts? Does either market have outliers?

5. Based on the boxplot from question 4, explain why you may have some concern using Student's *t*-distribution to test the hypotheses from question 1.

6.    Rather than using Student's *t*-distribution to estimate the *P*-value for the hypothesis test from question 1, we can use a randomization applet. Because the null hypothesis is a statement of no difference, the time a particular market exists is not dependent upon whether the market is a bear or bull market. Go to www.pearsonhighered.com/sullivanstats and open the applet *Randomization Test Market*. Or, enter the raw data into the StatCrunch spreadsheet. In StatCrunch, select *Applets > Resampling > Randomization test for two means*. Select "Bull" for Sample 1 and "Bear" for Sample 2. Click Compute!. Click "Data" under Runs to verify the sample mean number of months for each market is the same as you found in question 2.

7.    Click "1 time" and observe the window. The applet randomly assigns the market condition (bull or bear) to a time frame. Explain how this random assignment of outcomes is consistent with the statement in the null hypothesis. Click on "1" under Runs tô see the results of the random assignment. What is the mean length of a Bear Market for this random assignment? What is the mean length of a Bull Market for this random assignment?

8.    We want to obtain a measure of the likelihood of obtaining a sample mean as extreme or more extreme than the one observed under the assumption there is no difference in the length of time each market lasts. We do this by randomly assigning the market condition to the length of the market many, many times. Click "1000 times." What proportion of the randomizations is greater than the difference in sample means from question 2? What does this value represent?

9.    What is the shape of the difference in sample means? Does this suggest that Student's *t*-distribution may be able to be used to approximate the *P*-value for the hypothesis test from question 1? Use StatCrunch to approximate the *P*-value for the hypothesis test in question 1. Interpret the *P*-value.

10.    Based on your analysis does it appear to be the case that Bull Markets last longer than Bear Markets?

# Comparing Rental Car Prices

Is there a difference in the prices charged by Avis and Hertz for renting cars?  To answer this question, Ken decides to obtain a random sample of twelve prices charged by each company for a daily rental of various car types.

The data are shown in the following table.

| Airport/Car Type | Avis | Hertz |
|---|---|---|
| Denver/Midsize | 84 | 67 |
| Seattle/Compact | 90 | 86 |
| Seattle/Full Size | 111 | 110 |
| Chicago/Standard | 54 | 46 |
| Cleveland/Economy | 31 | 77 |
| Lincoln/Midsize SUV | 45 | 50 |
| Phoenix/Convertible | 144 | 140 |
| Los Angeles/Midsize SUV | 55 | 75 |
| Los Angeles/Midsize | 37 | 41 |
| San Francisco/Standard | 74 | 80 |
| Dallas/Premium | 70 | 92 |
| Orlando/Midsize SUV | 94 | 79 |
| Miami/Fullsize SUV | 171 | 183 |
| Boston/Mini Van | 75 | 89 |
| Philadelphia/Standard SUV | 134 | 136 |
| Minneapolis/Compact | 16 | 16 |
| Nashville/Standard | 55 | 63 |
| Las Vegas/Full Size | 58 | 69 |
| Salt Lake City/Standard SUV | 108 | 121 |
| St. Louis/Standard | 64 | 70 |
| New York/Fullsize | 70 | 74 |

*Source*: Expedia

1.    Explain why this is a dependent sample.

2.    Compute the sample mean price for Avis, $\overline{x}_A$. Compute the sample mean price for Hertz, $\overline{x}_H$. Compute the difference in sample means, $\overline{x}_A - \overline{x}_H$ .

3.    Compute the differences, $d =$ Avis $-$ Hertz for each location. Compute the sample mean of the differences, $\bar{d}$ . Compare this result to the difference in sample means from question 2. Explain why this result occurs.

4.    Suppose you were under the impression that the data was obtained via an independent sample. Test the hypothesis that there is a difference in the prices charged. What is the standard error of the estimate for this test? What is the $P$-value? How many degrees of freedom are there?

5.    Now, test the hypothesis that there is a difference in the price charged correctly, by treating the data as a dependent sample. What is the standard error of the estimate for this test? What is the $P$-value? How many degrees of freedom are there?

6.    The more degrees of freedom a test has, the better. Why? Explain why the loss in degrees of freedom is more than offset by the reduction in standard error.

7.    Why did pairing by location reduce the standard error? How might the standard error by further reduced?

# Differentiating Between Practical and Statistical Significance

> The Mathematics Department at Metropolitan Community College conducted a study in which 62 students who were enrolled in an online course in Intermediate Algebra were divided into two groups. Each group was administered five unit exams plus a proctored, comprehensive final exam. The students in Group 1 had Exams 1, 3, and 5 administered online and the exams were not proctored. Students were allowed access to notes, the text, and other learning aids. However, Exams 2 and 4 were administered as traditional paper and pencil exams that were proctored in a testing center. Students did not have access to notes, the text, or other learning aids. The students in Group 2 took all five exams online and had access to notes, the text, and other learning aids. None of the five exams was proctored. Both groups were administered a paper and pencil comprehensive final exam that was proctored in a testing center.
>
> *Source:* Flesch, Michael and Ostler, Elliot, "Analysis of Proctored versus Non-proctored Tests in Online Algebra Courses" *Math*AMATYC *Educator*- Vol. 2, No. 1, August, 2010.

1.      What is the response variable in this study?  Is it qualitative or quantitative?  What is the explanatory variable in this study? Is it qualitative or quantitative?

2.      The data below show the results from Test 2, which was proctored for Group 1, but not proctored for Group 2.

| | *n* | Mean | Standard Deviation |
|---|---|---|---|
| Group 1 | 30 | 74.30 | 12.87 |
| Group 2 | 32 | 88.62 | 22.09 |

a.      Conduct the appropriate test to determine if a statistically significant difference exists between the exam scores in the two groups.

b.      Comment on the standard deviations between the two groups.  Specifically, what do the standard deviations imply?

3.    The data below show the results from the final exam, which was proctored for both groups.

|  | $n$ | Mean | Standard Deviation |
|---|---|---|---|
| Group 1 | 30 | 65.70 | 27.66 |
| Group 2 | 32 | 74.84 | 20.10 |

a.    Conduct the appropriate test to determine if a statistically significant difference exists between the exam scores in the two groups.

b.    Discuss the practical significance of the results.  In other words, regardless of your results from part (a), do the data suggest there is a difference in the two groups?

c.    Suppose each group sample size is doubled.  Redo part (a) under this scenario.

d.    Discuss the difference between statistical and practical significance, in general.  Include in your discussion the role sample size might play in obtaining results that are not statistically significant, yet may have practical significance.

# Chapter 12

# Inference on Categorical Data

# Performing a Goodness-of-Fit Test

This activity will involve comparing the observed color distribution in a bag of M&M's to the expected color distribution that is advertised by the manufacturer. Do you think the advertised color distribution is accurate? How can you decide?

For this activity, you will need a bag of peanut M&M's in the traditional colors (no holiday or special packs).

1.    Open your bag of peanut M&M's and count the number of M&M's.

Number of M&M's in your bag:    _____

2.    The company that produces M&M products used to advertise on their website that the color distribution for peanut M&M products is as follows:

Brown   12%
Yellow  15%
Red     12%
Blue    23%
Orange  23%
Green   15%

Based on this information from the manufacturer, calculate the expected number of M&M's of each color in your bag. You will need to consider the total number of M&M's in your bag, along with the advertised proportions. Record these expected frequencies in the table below.

| Color | Expected Frequencies | Observed Frequencies | $\dfrac{(\text{observed} - \text{expected})^2}{\text{expected}}$ |
|---|---|---|---|
| Brown | | | |
| Yellow | | | |
| Red | | | |
| Blue | | | |
| Orange | | | |
| Green | | | |
| **Total** | | | |

3.    Now count how many M&M's you have in each color in your bag. Enter these observed frequencies in the table above.

4.    Are there any observed frequencies in the table that seem very different from the expected frequency? Explain.

5.     Write the hypotheses needed to test if your bag of peanut M&M's has the same proportions as those advertised by the manufacturer.

6.     Calculate the value of the chi-square test statistic.  Use the last column in the table to record your work.

7.     Write a conclusion for the test based on your value of the test statistic or the *P*-value.

# Testing for Homogeneity of Proportions

Have you ever read the information insert that comes with your prescription medication? Have you ever wondered if the side effects listed are more likely to occur with a higher dose of the medication than a lower dose? This activity will help you analyze the side effect rates for a drug and determine if there is a statistically significant difference in the side effect rates for different doses of the drug.

Suppose the following table lists the side effect rates for a particular drug, administered at two different dosages, and the placebo.

| Adverse Event | Drug (300 mg/day) ($n = 987$) | Drug (400 mg/day) ($n = 995$) | Placebo ($n = 992$) |
|---|---|---|---|
| Rash | 4.8% | 1.8% | 0% |
| Nausea | 1.6% | 3.6% | 0.6% |
| Agitation | 0.6% | 3.6% | 1.6% |
| Migraine | 0% | 3.6% | 1.6% |

1.    Calculate the observed and expected frequencies for each cell in the following table for the first adverse event.

| Adverse Event | Drug (300 mg/day) | Drug (400 mg/day) | Placebo |
|---|---|---|---|
| Rash | | | |
| No rash | | | |

a    Verify that the conditions are satisfied for the Chi-Square Test for homogeneity of proportions.

b.    Is there enough evidence to indicate that the proportion of subjects in each group who experienced rash is different at the 0.05 level of significance?

2.    Calculate the observed and expected frequencies for each cell in the following table for the second adverse event.

| Adverse Event | Drug (300 mg/day) | Drug (400 mg/day) | Placebo |
|---|---|---|---|
| Nausea | | | |
| No nausea | | | |

a    Verify that the conditions are satisfied for the Chi-Square Test for homogeneity of proportions.

b.    Is there enough evidence to indicate that the proportion of subjects in each group who experienced nausea is different at the 0.05 level of significance?

3.    Calculate the observed and expected frequencies for each cell in the following table for the third adverse event.

| Adverse Event | Drug (300 mg/day) | Drug (400 mg/day) | Placebo |
|---|---|---|---|
| Agitation | | | |
| No agitation | | | |

a    Verify that the conditions are satisfied for the Chi-Square Test for homogeneity of proportions.

b.    Is there enough evidence to indicate that the proportion of subjects in each group who experienced agitation is different at the 0.05 level of significance?

4.    Calculate the observed and expected frequencies for each cell in the following table for the fourth adverse event.

| Adverse Event | Drug (300 mg/day) | Drug (400 mg/day) | Placebo |
|---|---|---|---|
| Migraine | | | |
| No migraine | | | |

a    Verify that the conditions are satisfied for the Chi-Square Test for homogeneity of proportions.

b.    Is there enough evidence to indicate that the proportion of subjects in each group who experienced migraine is different at the 0.05 level of significance?

Is there enough evidence to indicate that the proportion of subjects in each group who experienced migraine is different at the 0.05 level of significance?

Calculate the observed and expected frequencies for each cell in the following table for the frontal adverse events.

| Adverse Event | Drug (300 mg/day) | Drug (400 mg/day) | Placebo |
|---|---|---|---|
| Migraine | | | |
| No migraine | | | |

a. Verify that the conditions are satisfied for the Chi-Square Test for homogeneity of proportions.

b. Is there enough evidence to indicate that the proportion of subjects in each group who experienced migraine is different at the 0.05 level of significance?

# Chapter 13

# Comparing Three or More Means

# Designing a Randomized Complete Block Design

How does design affect the flight of a paper airplane?  Explore this concept by comparing the flight distances of three different designs.  Since the type of paper could have an effect, use the same type of paper (for example, newspaper, brown shipping paper, typing paper, or other) and block by gender.

1.    Create a table to record your flight data based on gender (block) and plane design (treatment).

2.        Search the Internet for different designs of paper airplanes and choose the designs you wish to use.

          List the designs here and give them names so you can reference them in the table.

3.        Discuss how the model design and other variables might affect flight distance.  How can you account
          for these other variables?

4.        Fly each plane the same number of times and record the flight distances in your table.

5.        Was there a significant difference in flight distances for the different types of designs?  Explain.

6.        If you answered yes to the previous question, conduct Tukey's test to determine which differences in
          flight distance are significant using the familywise error rate of 0.05.

# Performing a Two-Way ANOVA

How does the type of paper affect the flight of a paper airplane?  In the first part of this activity, we looked at how the design of the plane affected flight distance while we blocked on gender.  Now we also want to know whether the paper type plays a role in paper airplane flight.  You will use three difference types of paper (for example, newspaper, brown shipping paper, typing paper) and perform a two-way ANOVA.

1.    Determine if there is a significant interaction between paper type and design.

2.    Draw an interaction plot of the data to support the results.

3.    a.    If there is not significant interaction, determine if there is significant difference in the means for the three plane designs.

      b.    If there is not significant interaction, determine if there is significant difference in the means for the paper types.

4.    a.    If there is significant difference in the means for the three plane designs, use Tukey's test to determine which pairwise means differ using a familywise error rate of 0.05.

      b.    If there is significant difference in the means for the paper types, use Tukey's test to determine which pairwise means differ using a familywise error rate of 0.05.

5.    Check the requirements for the model.  Are they all met?  If not, how will that affect your conclusions?

# Chapter 14

# Inference on the Least-Squares Regression Model and Multiple Regression

# Testing the Significance of a Regression Model

> This activity will revisit the arm length and height data collected from your class in Chapter 4. If you have already determined a linear model to relate forearm length and height, you can now test the significance of that model.

This activity will use the data on forearm length and height from the Chapter 4 activity. It will also use the linear regression model that was created in that activity.

1.    State the least-squares regression model that was obtained for the forearm and height data in Chapter 4.

2.    Verify that the requirements to perform inference on the least-squares regression model are satisfied.

   a.    Use an appropriate technology to construct a plot of the residuals against the explanatory variable (forearm length). Explain what you look for in this graph and what condition it helps you verify.

   b.    Use an appropriate technology to construct a normal probability plot. Explain what you look for in the normal probability plot and what condition it helps you verify.

3.    Test whether a linear relationship exists between forearm length and height at the 0.05 level of significance.

   a.    Write the hypotheses needed for this test.

   b.    Compute the value of the test statistic or obtain it from an appropriate technology.

   c.    Using either the classical or p-value approach, make a decision about the null hypothesis.

   d.    State the conclusion.

4.    a.    Construct a confidence interval for the slope of your regression line.

      b.    Write a sentence to interpret this confidence interval.

# Using a Randomization Test for Correlation

As stated at the end of Section 14.1, it is generally recommended that inference on the correlation coefficient be avoided because it is difficult to verify the requirement that the data come from a bivariate normal distribution. Instead, it is recommended that the equivalent hypothesis test of $H_0$: $\beta_1$ = 0 versus $H_0$: $\beta_1 \neq 0$ be conducted to determine if there is a significant linear relation between two quantitative variables.

Besides testing the hypotheses $H_0$: $\beta_1$ = 0 versus $H_0$: $\beta_1 \neq 0$ using the parametric methods presented in the text, we can also use a *permutation test*. A **permutation test** approximates the distribution of a statistic under the assumption the null hypothesis is true (in this case, the data are not linearly related) by randomly assigning a value of the response variable to a value of the explanatory variable and determining the correlation coefficient. This is repeated many times and the proportion of correlation coefficients that are as extreme as or more extreme than the observed correlation coefficient is determined. This value approximates the *P*-value of the distribution under the assumption the null hypothesis is true. The idea is that if the null hypothesis is true, then there is no relation between the explanatory and response variable, so it would be as if the value of the response variable is randomly assigned to the explanatory variable.

1.    Consider the following data, which represents the MRI count (a proxy for brain size) and IQ for 10 randomly selected female students.

| MRI Count | IQ |
|-----------|-----|
| 816,932 | 133 |
| 951,545 | 137 |
| 991,305 | 138 |
| 833,868 | 132 |
| 856,472 | 140 |
| 852,244 | 132 |
| 790,619 | 135 |
| 866,662 | 130 |
| 857,782 | 133 |
| 948,066 | 133 |

*Source*:  L. Willerrman, R. Schultz, J. N. Rutledge, and E. Bigler (1991). "In Vivo Brain Size and Intelligence," *Intelligence*, 15, 223-228.

a.    Draw a scatter diagram of the data and determine the correlation coefficient between the MRI Count, the explanatory variable, and IQ, the response variable.

b.    What do the scatter diagram and correlation coefficient imply?

2.    a.    Use the "sample from columns" option of a statistical spreadsheet to obtain 20 random samples without replacement from the IQ column. Determine the correlation coefficient between MRI Count and IQ for each of the 20 samples.

      b.    Now determine the proportion of correlation coefficients that are greater than the correlation coefficient in #1. This is an estimate of the $P$-value for the test $H_o : \rho = 0$ versus

            $H_1 : \rho > 0$, where $\rho$ is the population correlation coefficient between MRI Count and IQ. Explain why.

3.    If your statistical spreadsheet (such as StatCrunch) has a resample command, use it to obtain 1000 permutation samples and obtain the resample statistic. The screen below shows how to obtain the resampled statistic from StatCrunch.

What proportion of the resamples has a correlation coefficient greater than or equal to the observed correlation found in #1?

4.   a.   Go to www.pearsonhighered.com/sullivanstats and open the applet *Randomization Test Brain Size*. Or, open the *Randomization Test for Correlation Applet* on StatCrunch by selecting *Applets > Resampling > Randomization test for correlation*. Estimate the proportion of resampled correlations that are greater than the correlation found in #1 by running at least 3000 iterations of the randomization applet. That is, use the applet to test $H_o : \rho = 0$ versus $H_1 : \rho > 0$. What is the estimated *P*-value?

   b.   Now conduct the test $H_o : \beta_1 = 0$ versus $H_1 : \beta_1 > 0$. Compare the *P*-value from this test to the result from part (a).

5.   Consider the following data, which represents the estimating selling price (called the "Zestimate") and actual selling price of the home for a random sample of 10 recently sold homes in Oak Park, IL. Both variables are measured in thousands of dollars.

| Zestimate | Selling Price |
|-----------|---------------|
| 291.5     | 268           |
| 320       | 305           |
| 371.5     | 360           |
| 303.5     | 283           |
| 351.5     | 340           |
| 314       | 275           |
| 332.5     | 356           |
| 295       | 300           |
| 313       | 285           |
| 368       | 390           |

*Source*: www.zillow.com

   a.   Draw a scatter diagram and determine the correlation coefficient between the "Zestimate," the explanatory variable, and selling price, the response variable.

   b.   What type of relation appears to exist between the two variables?

c.    Go to www.pearsonhighered.com/sullivanstats and open the applet *Randomization Test Zillow*. Or, open the *Randomization Test for Correlation Applet* on StatCrunch by selecting *Applets > Resampling > Randomization test for correlation*. Estimate the proportion of resampled correlations are as extreme or more extreme than the correlation found in part (a) by running at least 3000 iterations of the randomization applet. That is, use the applet to test $H_o : \rho = 0$ versus $H_1 : \rho \neq 0$. What is the estimated *P*-value?

d.    Now conduct the test $H_o : \beta_1 = 0$ versus $H_1 : \beta_1 > 0$. Compare the *P*-value from this test to the result from part (c).